GREAT LAKES
JOURNEY

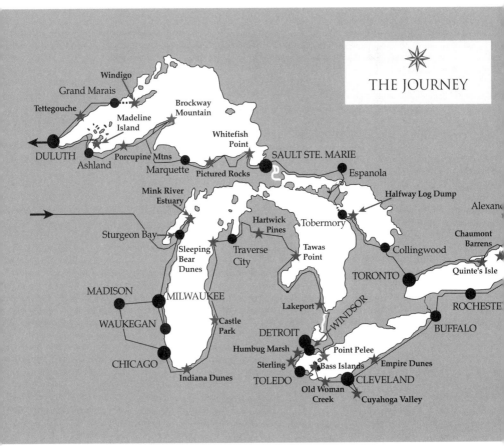

THE JOURNEY

Windigo
Grand Marais
Tettegouche
Madeline
Island
DULUTH
Ashland
Porcupine Mtns
Marquette
Brockway
Mountain
Whitefish
Point
Pictured Rocks
SAULT STE. MARIE
Espanola

Mink River
Estuary
Halfway Log Dump
Alexand
Sturgeon Bay
Hartwick
Pines
Tobermory
Chaumont
Barrens
Sleeping
Bear
Dunes
Traverse
City
Tawas
Point
Collingwood
Quinte's Isle
MADISON
MILWAUKEE
TORONTO
WAUKEGAN
Castle
Park
Lakeport
WINDSOR
ROCHESTE
CHICAGO
DETROIT
Humbug Marsh
Point Pelee
BUFFALO
Indiana Dunes
Sterling
Bass Islands
Empire Dunes
TOLEDO
CLEVELAND
Old Woman
Creek
Cuyahoga Valley

GREAT LAKES JOURNEY

A NEW LOOK
AT AMERICA'S
FRESHWATER COAST

WILLIAM ASHWORTH 1942-

WAYNE STATE UNIVERSITY PRESS   DETROIT

Great Lakes Books

A complete listing of the books in this series
can be found at the back of this volume.

Philip P. Mason, Editor
Department of History, Wayne State University

Dr. Charles K. Hyde, Associate Editor
Department of History, Wayne State University

Copyright © 2000 by William Ashworth
Published by Wayne State University Press
Detroit, Michigan 48201
All rights are reserved
No part of this book may be reproduced without formal permission
Manufactured in the United States of America
04 03 02 01 00   5 4 3 2 1

Library of Congress Cataloging-in-Publication Data

Ashworth, William, 1942–
Great Lakes journey : a new look at America's
freshwater coast / William Ashworth.
p. cm.—(Great Lakes books)
Includes bibliographical references (p.   ) and index.
ISBN 0-8143-2836-9 (alk. paper)
1. Great Lakes Region—Environmental conditions.   2. Great
Lakes Region—Description and travel.   3. Natural
history—Great Lakes Region.   4. Ashworth, William,
1942—Journeys—Great Lakes Region.   I. Title.   II. Series.
GE160.G75 A84 2000
333.91'63'0977—dc21
00-008766

All photos are by the author unless otherwise noted.

FOR JAMIE, BRYAN, WOLFGANG
AND NOW EMILY

Lake Superior lay before us. He who, for the first time, lifts his eyes upon this expanse, is amazed and delighted at its magnitude. Vastness is the term by which it is, more than any other, described. Clouds robed in sunshine, hanging in fleecy or nebular masses above—a bright, pure, illimitable plain of water—blue mountains, or dim islands in the distance—a shore of green foliage on the one hand—a waste of waters on the other. These are the prominent objects on which the eye rests. We are diverted by the flight of birds, as on the ocean.

Henry Schoolcraft, *Memoirs*, 1851

In recent times the Great Lakes have been less celebrated than other spectacular features of North America. Comprehension of the Lakes as a unit has been slow in coming, and this factor has contributed greatly to the ecological disasters which have enveloped the area. In many respects this is difficult to understand for they are truly awesome features. . . . It is perhaps the designation of "lake" that carries with it the diminutive concept of a watery interruption of the landscape. In fact, these waters dominate the surrounding land, bringing a marine element deep into the heart of the continent.

Tom Kuchenberg, *Reflections in a Tarnished Mirror*, 1978

All of the loons aren't out in the lake.

Peter Berryman, "Forward Hey," 1988

Contents

ACKNOWLEDGMENTS

A book is rarely a solitary project, and this one certainly wasn't. So many people contributed to it, in so many ways, that merely to list them all would require a separate volume nearly as large as the one you currently hold in your hand. What follows is thus necessarily incomplete—but that hasn't kept me from taking a stab at it.

I would like to begin by noting the very large debt my work owes to those Great Lakes residents whose expertise, local knowledge, and friendship I leaned on heavily during the long summer of 1998. A partial list would have to include (Lake by Lake):

Lakes Michigan and Huron: Merlin and Cathy McDaniel in Mequon, Wisconsin; Jane Elder, Bill Davis, and DuWayne Gebken in Madison; Bart and Helen Beavin in Naperville, Illinois; Bob Beltran, Kent Fuller, and John Wasick in Chicago; Amy Blossom, Brad Galusha, and their children and extended family in Castle Park, Michigan (Amy also put up with my absences and occasional detached befuddlement at work, in her other role as my coworker at the Ashland Public Library back in Oregon); Max Holden at Sleeping Bear National Lakeshore; and Ted Sobel on board the *Chi-Cheemaun.*

Lakes Erie, Ontario, and St. Clair: John Hartig of the International Joint Commission in Windsor, Ontario; Eddie and Ricki Herdendorf, John Hageman, John and Phyllis Crites, Melissa Haltuch, and Dave Moore at Put-in-Bay, Ohio; Bob and Maria Bertram in Lorain, Ohio; Ed and Anna Fritz in Cleveland; Christopher Densmore of the State University of New York at Buffalo (and Buffalo Friends Meeting); Bill

"The Guard" Bogacki at the Tifft Nature Preserve; Andy Frank and Reg Gilbert at Great Lakes United; and Joe DiPinto at the Great Lakes Program office at SUNY-Buffalo.

Lake Superior: Dorette Roach at the Whitefish Point Bird Observatory; Bill and Glenda Robinson in Marquette, Michigan; David Carper and Cara Norland on the beach just outside Porcupine Mountains State Park; and Lauri and Dan Bliss in South Range, Michigan.

In addition to these, there were numerous helpful individuals whose names I either learned and quickly forgot or simply never learned at all. These include state, provincial and national park staff persons at all levels in both countries; motel and restaurant owners and employees all over the Basin (but especially in Traverse City, Michigan, and Grand Marais, Minnesota); the Door County constable we spoke with at Cave Point, Wisconsin; the retired Canadian civil servant we ran into on the Isle of Quinte in Ontario; the ubiquitous tourist information booth attendants, especially those in Sault Ste. Marie, Michigan, and in Colbourne and Blind River, Ontario; the young woman from the staff of the *Chi-Cheemaun* in Tobermory, Ontario, who ran my key back to the motel when early-morning fog (intracranial) caused me to forget to turn it in; the attendant at the Superior Water-logged Lumber gift shop in Bayfield, Wisconsin; and the passengers and crew of the wave-tossed *Wenonah* between Grand Portage, Minnesota, and Isle Royale National Park.

Rod Badger in Ashland, Oregon and Larry Chitwood in Bend, Oregon—my Great Lakes companions from 1983—each met with me several times around the present trip to help me prepare beforehand and decompress afterward. Larry and his wife, Karen, also provided lodging and R&R on the trip to and from the Lakes, as did my sisters Lillian and Judith Ashworth in Pullman, Washington. Lodging in the Great Lakes Basin was kindly provided for us by Merlin and Cathy McDaniel; Bart and Helen Beavin; Eddie and Ricki Herdendorf; and Ed and Anna Fritz. Larry Calkins, Julie Riker, and Karen Taylor took turns watching our house and feeding our two cats while we were gallivanting about the landscape two thousand miles or so away— something the cats probably appreciated even more than we did.

Research queries from Oregon both before and after the trip were handled patiently and expertly by (in alphabetical order) Jim Bredin of the Michigan Office of the Great Lakes; Curtis Curtis-Smith at Western Michigan University; Robert Fudge of the National Park Service in Washington; Smitty Parratt at Isle Royale National Park; Charlotte Read of Save the [Indiana] Dunes; Stone Lab (Ohio) director

Jeff Reutter; Brian Robinson and Karen Wallace at the Tifft Nature Preserve; Russell Utych at the Whitefish Point Bird Observatory; and Margaret Wooster, executive director of Great Lakes United. Bob Beltran, Amy Blossom, Bill Bogacki, Jane Elder, Andy Frank, Ed and Anna Fritz, Kent Fuller, Dorette Roach, Russell Utych, John Wasik, and Margaret Wooster all reviewed parts of the manuscript prior to publication; DuWayne Gebken, John Hartig, Eddie Herdendorf, and Bill and Glenda Robinson reviewed all of it. (All comments I received were friendly and constructive, and all were appreciated, even those few—few from *my* standpoint, at least—that I did not choose to follow). Max Gartenberg handled the business details of the contract in his usual competent manner, and managing editor Kathy Wildfong at Wayne State University Press encouraged the book and its author through many moments when both seemed to be flagging.

Finally, I need to acknowledge the biggest contribution to the book other than (perhaps) my own: that of my traveling companion, research assistant, chief editor, logistician, lover, and wife of thirty-two years, Melody James Ashworth. I have thanked her privately many times; I am very happy to seize the opportunity to thank her publicly. I might have been able to do this book without her. I am certainly glad I didn't have to try.

Solutions to the problems facing the Great Lakes cannot be imposed by outsiders viewing the place from Oregon (or anyplace else); they can only be grown from within, by Great Lakes residents themselves. Mindful of this, I would like to articulate once more one of the principal themes of this book: not in my words, but in those of a third-generation Great Lakes dweller, Dorette Roach. The passage comes from a recent e-mail message from her, quoted with her permission:

> I love the Whitefish Point area and can understand people wanting to visit there. It's just that there seems to be a certain amount of willful ignorance in tourists, who trample anything in their pursuit of recreation. They do not realize the irony that the things that they come here for are the things they change by being here. When they are faced by this beautiful lonely shore they say "What do you DO here?" "Where is the closest McDonald's?" "How come none of these motels have a pool?" Complaining about the lack of amenities will have one effect: someone will supply their demand. Where will the tourists go when this place is just like the places they left? I have no solutions to this problem. I know that our livelihoods depend on tourist dollars. I just want you to know that my frustration is not unfriendliness.

ACKNOWLEDGMENTS

Credit for this book really belongs to Dorette—and to Bill Robinson and Eddie Herdendorf and Jane Elder and Ted Sobel and all the others whose names and ideas appear in it. I am pleased to have been the conduit for them.

Ashland, Oregon, September 13, 1999

PROLOGUE:
PULASKI SQUARE

FATE RARELY CALLS FOR AN APPOINTMENT: it prefers to sneak up and smack us unaware. Mine caught up with me in Cleveland, Ohio, on a cold, gray, dismal March morning in 1981.

I had come to Cleveland to do an interview on—well, the subject no longer seems to matter very much. The flight I had booked arrived at Cleveland Hopkins International Airport on Sunday morning, and the man I had come to see was leaving for church. There were several hours to kill. It was March, it was cloudy, and it was Cleveland. Not, I thought, a particularly hopeful prospect.

I took the train downtown, to the bowels of Cleveland's fusty old beaux arts Terminal Tower. I stowed everything but my camera in a coin-op locker and ascended several dreary flights of broad marble stairs to street level. Outside, wind gusts rattled vagrant newspapers in Public Square. The sky was gray. The gray had bled onto everything else in sight. I could feel it running along my skin, testing my pores, trying to get inside.

I walked east and north up Rockwell and Saint Clair and Sixth, aiming vaguely toward city hall. I crossed Lakeside to Pulaski Square. An ancient Polish cannon squatted near the far edge of the square, pointing north. I wandered over to it, peered into the barrel, sighted along it.

There was the Water.

It is one thing to read of fresh water that stretches like an ocean to the horizon: it is quite another to actually *see* it, beyond the colorless winter weeds at the bluff's edge, beyond the cut-up, mangled waterfront, a restless blue plain—blue, though the sky was gray—running north, running east, running west, beyond sight, beyond knowledge, beyond imagination. Gulls swooped above it, crying. I felt a click and shift in my life, and I knew it would never be the same.

Since that moment in Pulaski Square I have visited the Great Lakes many times. I have toured their coastlines, walked their woods and beaches, waded in their waters, and sailed upon them in vessels ranging from canoes to ocean liner–sized ferries. I have come to know them in storm and sunlight, by day and night, in fog and smog and morning dew. I have photographed them, researched them, and written about them. And I have watched them change.

It is the changes I wish to address in this book. There have been many of these during my long acquaintance with the Lakes: changes in water transparency, changes in contaminant levels, changes in species composition, changes in shoreline development. Changes in attitudes. Governments have changed the way they manage the Lakes; scientists have changed the way they study them. Parks have changed the way they protect them. About the only thing that hasn't changed is the basic shape of the Lakes themselves, and there are detail differences even in that.

We North Americans are uncomfortable to the point of irrationality with change. This applies equally well to developers, whose madness is to try to deny the existence of change, and to environmentalists, whose madness is to try to stop it. But change is funny: it gives the illusion of being easily stopped or denied, but it is actually not possible to do either. You can manage it, to an extent, but you cannot control it. Like fate, it will sneak up on you.

In the summer of 1983, still in the first flush of the epiphany that had struck me in Pulaski Square, I took a long trip by car through the Great Lakes Basin. Coming in through the Door—Wisconsin's Door Peninsula—I looped south around the bottom of Lake Michigan and loitered up its eastern shoreline to the Straits of Mackinac. I skirted Lake Huron's North Channel, passed through Sudbury, headed south through French River and Parry Sound to Toronto. Swung around the eastern end of Lake Ontario, through the Thousand Islands. Came on west through Rochester and Buffalo and Cleveland and Toledo; north through Detroit and Saginaw and Alpena and St. Ignace; then west again, through Pictured Rocks and Marquette and the Keweenaw and the Apostles to Duluth and the North Shore, which I followed as far as Thunder Bay, Ontario. When I got home to Oregon I put what I had found into a book. It was called *The Late, Great Lakes,* and it told of the Great Lakes' past and present and some of their uncertain future.

In the summer of 1998, fifteen years after the first trip, I toured the Basin once more. I followed much the same route, went to most of the same places, and talked to many of the same people. This book is a report of what I found the second time around.

BASELINE DATA. CHANGE CANNOT PROP-
erly be understood outside the context in which it occurs.

Together, the five Laurentian Great Lakes and their connecting channels form the largest reservoir of fresh surface water on the planet. Measuring 1,100 miles from end to end and 500 miles from top to bottom, they sprawl over nearly 100,000 square miles—more than all the New England states put together—and contain nearly 5,500 cubic miles of fresh water, just under 20 percent of the total world supply. (Those who enjoy mundane comparisons might note that this is enough to flush an old-fashioned five-gallon toilet approximately 22 trillion times.) The United States has over 4,000 miles of coastline on the Great Lakes, more than it does on the Atlantic Ocean and the Gulf of Mexico combined: Michigan's coast alone extends for more than 2,232 miles, giving it more shoreline than any other state in the Union except Alaska. The U.S.-Canadian border passes down the center of the Lakes for a thousand miles. We think of this border as a line on the land, and of course it is; but along much of that line, a citizen standing in one nation and looking toward the other will see nothing but restless blue water.

Geologically, the Lakes are extremely young. They began peering out from beneath the retreating edge of the continental glacier barely twelve thousand years ago, and the lands around them continue to bear the marks of ice—scarred and polished bedrock, huge moraines, numerous bogs and kettles. A confused welter of little rivers flows hither and yon across a landscape so new that it has yet to establish its permanent drainage pattern. So new, in fact, that it has yet to establish its permanent elevation. Twelve thousand years ago the continental surface here was depressed beneath the massive weight of two vertical miles of frozen water, and ever since the glacier melted away the land has been slowly rising back to its pre-ice height, a process known to geologists as *isostatic rebound*. Old beach lines and wave terraces, some of them hundreds of feet above current Lake levels, show the past extent of this rise. The process continues today, tipping the Lakes slowly toward the south. The water level is creeping up on the south shores of all the Lakes; it is creeping down on their north shores. The movement is measurable only in millimeters per year, but it is measurable—rapid enough that, if you are going to state the elevation of a point in the Great Lakes Basin with any real accuracy, you must give a year as well as a height.

Since the oldest human artifacts found in North America date from before 12,000 years ago, it is safe to say that human cultures witnessed the birth of the Great Lakes and all the subsequent changes in their geography, from the Chippewa-Stanley Low-Water Stage 9,500 years ago to the Nipissing High-Water Stage five millennia later. The written account, however, does not begin until August 1, 1615. Toward evening on that day the French explorer Samuel de Champlain, a son of a sea captain but a confidante of kings, drifted in awe out of the mouth of the French River onto the oceanic expanse of Lake Huron, which he called *La Mer Douce*—the Sweet Sea. By October of that year Champlain had also canoed much of the length of Lake Ontario (or perhaps across it: the record is unclear). Lake Superior was first written up by Europeans in 1622; Lake Michigan, in 1634. Oddly, Lake Erie—the oldest and southernmost of the Great Lakes, and today the most heavily urbanized—was the last to come onto the maps: its existence wasn't confirmed to European scholars' satisfaction until 1669.

Though the Great Lakes have served Western culture as an avenue for transportation into the heart of North America from the beginning of our presence here—there was a sailing vessel on Lake Erie as early as 1679, just ten years after the Lake's official "discovery"—serious development didn't get under way until the middle of the nineteenth century. As late as 1820, while Lewis Cass was governor of Michigan Territory, so little was known about the Lakes and the lands they lapped against that one of the governor's official acts was to lead an exploring party into the wilderness to find out just what it was that he was supposed to be governing. After that, however, things sped up considerably. Chicago, a village of "ten or twelve dwelling houses, with an aggregate population, of probably, sixty souls" when Cass's party visited it, had exceeded a million people by the end of the century. Detroit was approaching the 300,000-person mark by this time; Milwaukee had passed it. Toronto was at 208,000 and rising. Industry was well established in the region: iron and copper mines riddled the mountains around Lake Superior, steel was being forged in Cleveland and Chicago and Hamilton, and Henry Ford was starting to build automobiles in Detroit. The Big Cut had come, and the trees had gone. Canals had been constructed around all of the navigation obstructions on the connecting channels—including, notably, Niagara Falls—and most harbors had been dredged and provided with breakwaters. Barely fifty years removed from the last days of the voyageurs, the Lakes had entered the modern era.

The first signs of stress were visible as early as the 1890s, when the flow of the Chicago River was reversed to clean up a serious fecal-contamination problem on the beaches of the upper-crust Chicago residential area known as the Gold Coast. Not a great deal of attention was paid, however, until the 1950s. That was when sea lamprey numbers suddenly exploded, decimating whitefish and lake trout populations and sending Great Lakes fisheries into a precipitous decline. The lampreys had barely been brought under control—sort of—when Lake Erie went eutrophic (a condition widely trumpeted as "dead," though eutrophic lakes, far from being moribund, are actually too alive for their own good). The spectacle of a 10,000-square-mile body of water turned stinking green galvanized the fledgling environmental movement, which gained even more ammunition when the Cuyahoga River—flowing into Lake Erie through the heart of Cleveland's heavily industrialized Flats—caught fire and burned in the late 1960s. Again. At about the same time, the alewife—a four-inch-long invader from the North Atlantic which had entered the Great Lakes by way of the shipping canals—became the most prolific fish species in the Great Lakes, accounting at one point for more 90 percent of the total piscine biomass in Lake Michigan. Masses of dead alewives began washing up on beaches. Bottom samples taken from harbors and river mouths contained high concentrations of toxic and carcinogenic chemicals: 74,000 parts per million at Toledo, 325,000 ppm at Gary, a whopping 500,000 ppm at Waukegan. In the middle of Isle Royale, in the middle of Lake Superior, fish from wilderness lakes showed whole-body PCB concentrations well above the level considered safe for human consumption. Most of the 24 million residents of the Great Lakes Basin drew their drinking water from the Lakes, and a fairly high proportion of them regularly ate Great Lakes fish. What, they began to wonder, were they doing to themselves?

By the time I made my first extensive Great Lakes tour in 1983, answers seemed to be becoming known.

When I set out again in the summer of 1998, answers still seemed to be becoming known. But sometimes they were not the same answers.

PEOPLE CHANGE, AS WELL AS PLACES. THE filters through which I observed the Lakes in 1983 were quite different from those through which I observed them in 1998. In the interest of objectivity, I should stipulate a few of those differences here.

The biggest difference is also the most subtle: I am fifteen years older. Age has brought a certain slowing down: fewer miles walked during the day, fewer campgrounds and more motels at night. It has also—dare I say this?—brought at least *slightly* more wisdom. At forty-one I was an active environmentalist—a member of the Sierra Club's Oregon Chapter Executive Committee, which I had helped found and on which I had served as vice-chair, policy coordinator, and secretary. At fifty-six I am still concerned about the state of the environment, but I have forsaken activism in favor of pragmatism. I have come to mistrust preservation, which seems to me to partake of the same faulty worldview as does development: humans as fundamentally separate from nature, one the conqueror, the other the conquered. Both sides seem to view the human race as a monkey wrench, either delicately adjusting the machinery or being flung into it—a distinction which no longer appears to me to hold any real significance. At any rate, that is the view from my current set of biases. I am no longer a member of the Sierra Club.

A second major difference in the way in which I viewed the Lakes fifteen years apart lies in the nature of the company with which I viewed them. This, too, is related to aging, though indirectly. In 1983 I made the trip with two friends, geologist Larry Chitwood and organic chemist Rod Badger. Larry took the first half, Rod the second: they changed in the middle, at Detroit, rather like runners in a relay handing off a baton. In 1998, by contrast, I did what I had wanted to do the first time but could not, due to the ages of our children: I took my wife along. Melody is a biologist by background and a medical technician by profession, and that gave me—as had the geologist and the chemist—a scientifically trained mind to bounce ideas and observations off of. Unlike the geologist and the chemist, though, she is also my wife, and even after thirty-one years of marriage traveling with her retains aspects of an extended honeymoon. There is something profoundly different between watching a sunset with a friend—even a close friend—and watching it with your lover. I have no doubt that this altered my perceptions, at least slightly, all along the route.

A third change was the removal of novelty. Except for that glimpse in Cleveland, I had not seen the Lakes at all before the 1983 trip. I had traveled fairly extensively around the rest of the continent, but it had all been by public transportation: I hadn't driven east of Idaho. By 1998, though, I had crossed the country several times by car—and most of those trips had an extensive Great Lakes component. The Lakes, and the routes to them, were old friends. Again, I have no doubt that this affected my perceptions.

In 1983, I took my trip in a green 1974 Subaru station wagon. We folded the backseat down and filled the space from the front seat to the tailgate with baggage and camping gear and still felt cramped in the tiny car—even more so during the latter part of the trip, after we had lashed a seventeen-foot canoe to the top and added paddles and life jackets to the load in back. The sound system was a cheap portable tape player wedged between the backrests of the two front bucket seats at roughly shoulder level; I took it out at each stop to record interviews. For air-conditioning we rolled down the windows and drove with clothing wet down at rest stop water fountains.

The 1998 trip was made in a grey 1991 Ford Escort four-door hatchback—small but not really cramped. The car had cruise control, electric door locks, an adjustable steering wheel, and a decent-quality sound system. We had the air-conditioning rebuilt just before leaving home. There are some parts of an adventure that you want to duplicate the second time around. There are other parts it is much better to do without.

I
EASTWARD

It takes five days to drive from Oregon to Lake Michigan. There are people I know who can do it in three, but that requires traveling all night as well as all day, taking turns behind the wheel. We tried that once a few years ago, and you know what? It gets dark at night. There are few things more frustrating to a true, dyed-in-the-wool scenery addict than being surrounded by unfamiliar countryside you can't see. I have been accused, with some justification, of being a destination-oriented traveler—one who is not satisfied with a journey until the end point is reached. But of what use is an end point if it cannot be placed in the context of what is passed through to get there?

So we took five days: north first to Spokane, Washington, and then east through Idaho, Montana, North Dakota, and Minnesota, crossing into Wisconsin early on the fifth day. For the first two days we traveled among mountains, through the Cascades and the Rockies and over the rough, dry, scabbed surface of the great trough between them called the Columbia Basin. Toward the end of the third day, in eastern Montana, the mountains began to peter out, and after a last gasp of vertical relief in the Badlands around the Montana/North Dakota border—where we stopped for a sunset hike among brilliantly colored buttes—they quit altogether and the plains closed in. Those who live along the rumpled edges of the continent have little concept of what it is like out in the flat center, where mountain-building is only a dim geologic memory from the early days of the planet. We drove endlessly eastward toward an endlessly receding horizon, and the perfect dome of sky above us became a prison.

Near Eau Claire, Wisconsin, on the fifth day, the Great Plains drew at last to an end. We picked up Route 29 and came across the center of the Badger State to Green Bay. Wisconsin is a land of long green swells and little rivers which the highway builders have mostly ignored, so the road went endlessly slightly up and slightly down as it cut across the grain of the landscape, a style sometimes referred to cynically by civil engineers as "ridge and bridge." Between Wausau and Shawano the road was under construction, and we found ourselves in a seemingly endless series of pilot car–led caravans, chugging in lockstep past a seemingly endless series of closed and barricaded rest areas as the sun beat down and the hydraulic pressure in two human bladders built up. In the middle of that stretch, on a low, hardwood-clad ridge that was visually indistinguishable from its neighbors, we crossed from the little valley of the Plover to the little valley of the Embarrass and so entered the Great Lakes Basin. At least that is what my map says today. I am sorry to tell you that, beyond the pilot cars and the closed rest areas and the squirming, accelerator-tapping need to find an open one, I have no memory of that passage at all.

Things got better after Shawano. The Wisconsin sky was early-summer blue and dotted with clouds; the land was green and fair and seemed to be yearning toward the Lake. Shortly before 1:30 in the afternoon we crossed the last bridge and topped the last ridge and came down upon the Fox River, and Green Bay.

II
THE DOOR

HIGHWAY 29 INTERSECTED U.S. 41 ON THE
western outskirts of the city of Green Bay, in the middle of a wetland
stretching up from the bay along little Duck Creek. We turned onto 41
and drove six miles south to De Pere, then crossed the Fox and came
downriver on the east side, staying as close as possible to the water. At
Voyageur Park we pulled off and went down to the riverbank, across a
lawn littered with the big green droppings of Canada geese. The park
was just downstream from a large Nicolet Paper plant; another large
paper plant loomed on the far bank, a half mile or so downstream. The
Fox glided between them, broad and inviting. This is the largest river
flowing into Lake Michigan, and it would be a major recreational asset
to the communities along its banks if someone would just bite the
bullet and clean it up. People were using it anyway—small inboards
were going upstream and down past the park, and there was a water
skier out—but many more would surely use it if they could trust it.
Right now it is just not trustworthy. Decades of abuse have left it one
of the filthiest rivers in the Great Lakes Basin. Standing there amidst
the goose turds, watching the boaters and the skier, I couldn't help
feeling I was in the cleaner of the two spots.

In downtown Green Bay we picked up Highway 57 and followed it
up the middle of the Door Peninsula to Potawatomi State Park, where
we had camp set up by midafternoon. A hundred yards or so south of
our campsite, a low band of cliffs marked an old shoreline of Sturgeon
Bay; roughly the same distance to the north, small waves smacked
against the bay's current shore. Dolomite boulders scarred by the
claws of ancient glaciers lay scattered around us through a woodland
of white and yellow birch. Dolomite is why the Door is here. It is a
variety of limestone; but whereas most limestone is largely calcium
carbonate—a relatively soft and friable substance—dolomite is half
magnesium carbonate instead, and it has the strength and toughness
of granite. Because of this it was able to resist the glaciers that ground

most of the rest of the upper Midwest flat. Eighty miles long and twenty-five miles wide at the base, the Door is shaped like a giant neolithic dagger with its point aimed at Michigan's Garden Peninsula across the narrow passage known to the voyageurs as the *Porte des Morts*—the Door of the Dead. The Garden, too, is made of dolomite. It and the Door are part of a great ring of the stuff, more than 450 miles in diameter, which humps through much of the Great Lakes Basin like the backbone of an immense buried snake. Where this ring of stone passes between Lake Erie and Lake Ontario, the gathered waters of the four upper Great Lakes pour over it in a stupendous fall that—quite aside from thrilling honeymooners for two hundred years—has given its name to the entire circular feature. It is called the Niagaran Cuesta.

To the west of the Door lies Green Bay, a Lake Michigan embayment which, at 120 miles in length and 20 in width, dwarfs most of the world's large lakes. To the east lies Lake Michigan itself, the ocean of fresh water to which Green Bay is merely a minor appendage. Between them the flat, slightly tilted surface of the dolomite—low near the Lake, high near the bay—reaches as much as three hundred feet above Lake level. Potawatomi is on the bay side, but it is not that high. Halfway up the peninsula a freshwater fjord, Sturgeon Bay, slices most of the way across from Green Bay to the Lake. Humans finished the job in 1878, cutting a canal along the route of an old Indian/voyageur portage trail and converting the upper half of the peninsula into an island. Potawatomi's Daisy Field Campground lies along the south shore of this fjord-*cum*-canal, near the western end where God dug it, nestled onto a shelf cut back into the bluffs by the Lake when its level was somewhat higher than it is today. This was my third stay at Potawatomi. I spent my first night on the Great Lakes here in 1983, with Larry Chitwood; Melody and I also camped at Potawatomi on our first family trip to the Great Lakes, the time we brought our oldest daughter to the Midwest to begin her freshman year in college—Northland College, in Ashland, Wisconsin—in 1987.

Green Bay wasn't green, which really wasn't much of a surprise. The shallow bay comes by its name honestly—algal blooms have turned it green numerous times, off and on, in the roughly 370 years since the French explorer Jean Nicolet became the first white-skinned tourist to come here—but the blooms only happen in low-water years, and there haven't been many of those lately. We have also stopped augmenting the bay's natural eutrophication rate by dumping nitrates and phosphates into it, but this probably isn't as significant as the run of high water. This is, after all, the homeland of the Winnebago tribe, a people whose name translates roughly as "stinking water." I don't want to entirely discount the impact of cultural eutrophication, but we didn't cause most of the bay's nutrient overload, and we can't take much credit for the solution, either.

Which is not to say either that the bay is clean or that we are innocent. In fact, Green Bay is pretty dirty, and we are pretty culpable. We are no longer dumping much into it, but most of what we dumped earlier is still there. These are not nitrates and phosphates we are talking about, now, but toxic chemicals—PCBs, dioxins, and others, primarily members of the large group of compounds known as halogenated hydrocarbons. They are no longer spilling massively from factory pipes, but they reside in the bottom sediments, on the floor of the Fox River and in the gooey underwater delta the river spews into the south end of the bay. Before the Fox can enter Green Bay (the bay) it has to pass through Green Bay (the city), and in the city it runs a crowded gauntlet of industry, primarily paper mills. There are more paper mills per river mile on the lower Fox than on any other waterway in the world. Back in the days before anybody important cared very much, these mills dumped huge amounts of halogenated hydrocarbons into the river. Most of this torrent of persistent poisons ended up on the bottom, mixed into the sludge—a very descriptive word. We now know methods for getting the poisons out again, but it would cost the annual gross national products of several moderate-sized European countries to do the job, and so far no one has stepped forward to take responsibility. The river bottom was a reeking nightmare when I came through in 1983. It is less so today, but, as with the lack of algal blooms, we can't take a great deal of the credit. Most of the improvement has been done by Ma Nature, slowly covering the old toxic sediments with newer, less toxic sediments. Just like we've been doing. Sweeping the problem under the rug.

This is probably a good place to put in a first word about the twinned Great Lakes alphabet-soup programs known as AOCs and RAPs, because the Fox is a splendid example of what has gone wrong with them. AOC stands for Area of Concern, a designation used by the International Joint Commission on Boundary Waters (IJC)—the U.S.-Canadian agency whose primary bailiwick is the Great Lakes. Areas of Concern have been around since well before my first trip to the Lakes, but the term was legally meaningless before 1987. In that year—a year after publication of *The Late, Great Lakes*—the U.S.-Canada Great Lakes Water Quality Agreement underwent one of its periodic convulsive revisions, and the AOCs were given some teeth. The teeth came in the form of the RAPs—Remedial Action Plans. The revised Agreement designated forty-three specific AOCs around the Lakes, and it pledged the Parties to the Agreement—the two Great Lakes nations—to put a RAP in place for each AOC, detailing precisely how and when cleanup would take place. The RAPs were to be prepared through consensus, bringing environmental groups, industries, and government agencies to the same table, so that the emerging plan would have the support of as many of the stakeholders in the AOC as possible. Where "Potentially Responsible Parties"— PRPs, or, as they were rapidly nicknamed, "Perps"—could be identified, they would be required to pay the cleanup costs. Otherwise, government would foot the bill. Reports would go out regularly to the IJC, which was supposed to oversee the process and delist each AOC as its RAP was carried to completion.

The RAPs were a brave step forward, and in some places they were remarkably successful. The Fox wasn't one of these. Politically hamstrung, starved for lack of funds, the Fox River RAP went mostly nowhere. With government belt-tightening withdrawing support for the RAP process nearly everywhere around the Lakes these days, it doesn't look as though it ever *will* go anywhere. Wisconsin activists have largely given up on pursuing it, and are petitioning the Environmental Protection Agency to declare the entire lower twenty-nine miles of the river a Superfund site instead. That might mean replacing one dubiously successful program with another one just as dubious, but at least it won't require jumping through the hoops of an international treaty.

We will have much more to say about RAPs as time goes on, and most of it—happily—will be better news than this.

Potawatomi State Park abuts the town of Sturgeon Bay, a lovely old portside community that spills up and over the bluffs along both sides of its namesake body of water. Two bridges—one low and old, the other high and new—connect the halves of the town across the fjordlike bay. We picked up egg salad sandwiches and stuffed tomatoes at a small supermarket deli a short distance north of the low, old bridge and headed for Cave Point, on the Lake Michigan side of the peninsula. The small county park there was the scene of one of the more memorable picnics Larry and I shared, back in '83; I had visited it with the family in '87, as well. The associations were all positive. It would be a good first spot to really touch the Lakes, this time around.

It was nearing 5 P.M. and the light was turning golden-delicious as we drove down the long, narrow tunnel of trees leading into the park. The parking lot was half full, but there were only a few people in sight. I leaped from the car and, laden with cameras as I was, practically ran across the green lawn to the edge of Cave Point's low dolomite cliffs. The Lake stretched out forever, and the gulls played, and the waves smacked into the caves that lace the cliffs and give the point its name, and all was right with the world. We scrambled about the ledge that hugs the base of the cliffs near the northernmost cave. Gulls scrambled with us, begging for handouts. I looked for zebra mussels, but I didn't find any. These thumbnail-sized European mollusks were first identified in the Great Lakes system in 1988, in Lake St. Clair, where they had evidently been discharged in a load of ship's ballast water; since then they have spread throughout the system. Voracious filter feeders, they have dramatically altered the ecology of the Lakes, competing with other plankton eaters for food and driving the native mussels to near-extinction. I had fully expected zebras to be the most noticeable change in the Lakes, and was vaguely disappointed to find that Cave Point couldn't provide me with any. Maybe next time.

The low surf occasionally washed over the ledge, wetting our shoes. Infinite horizons opened to the east, over the big water.

As we were climbing the fifteen feet or so up from the ledge to the park lawn we came across a camera. It was a 35mm point-and-shoot with film in it, and it had been tucked into a niche in the cliff—obviously for safekeeping, just as obviously forgotten. There was no one around to whom it seemed to belong.

We briefly considered turning it in at the Sturgeon Bay police station. That seemed impractical: probably the person who lost it would never think to check there. In the end we decided the best thing to do would be to leave it in the park, in plain sight. We put it on a picnic table. Melody wrote a short note explaining where it had been found and left that on the table, too, weighting it down with the camera.

Straightening up from putting my own cameras in the backseat of the car a few minutes later, preparing to climb into the driver's seat, I noticed a slender, chino-clad man of about sixty reading Melody's note. He was holding the camera in his hand, and his expression was puzzled. I walked back to the table.

"Is that yours?" I asked, pointing.

He shook his head. "No," he said. "Looks like somebody found it down over the side."

"We did. My wife wrote that note. Nobody here seems to claim it. Do you think we should turn it over to the police?"

The man grinned. "You just did," he said. "I'm the local constable. I just live down the road a little ways. Came up here to play with my grandson." He threw a thumb over his shoulder at a boy of about ten who was climbing down from the passenger side of a large four-wheel-drive pickup. The boy waved. I waved back. The constable frowned at the camera again, holding it at arm's length in both hands. "But what am I gonna do with this thing?" he wondered, to no one in particular.

"We were just going to leave it there with that note."

"Too many city people around." He suddenly seemed to realize we might be city people. "You from around here?" he asked.

"No, we're from Oregon."

"That your Escort? I saw the plates. Is this your first visit to Door County?"

"Actually," I said, "I was here fifteen years ago."

He nodded. "Fifteen years ago, maybe you could have left it on a table. But things have changed." He broke off as a car drove up.

"Maybe they came back to look for it," he said hopefully, and he went over to speak with the driver.

He was back in a moment. "No," he said, "they just got here. You know, I think what I'll do is take it back to my house. I'll leave a note here on the table telling where it is. They come looking for it, maybe they'll find the note. Meantime, if you run into anyone looking for a camera, tell 'em where it is." He described in detail how to get to his house.

"We were just leaving," I said.

"Well, if you see anyone, let 'em know. It's just down the road."

As we left the park he was sticking his note to the table in place of Melody's, impaled on a twig pushed down between two of the planks forming the tabletop. It was 6:30 P.M. We drove back to Potawatomi through the clear evening air and climbed the observation tower on the hill above the campground, looking down to the north on Sawyer Harbor with its picture-perfect scattering of islands. Light, still, was nearly perfect. Back in camp we walked the beach for a short time along Sturgeon Bay, then came home to our campsite. The dark came down. The dolomite boulders glowed faintly in the dusk, like rock ghosts.

The Door is as lovely as I remember, and it is still primarily agricultural. But the constable was right: a problem is developing. Tourists were already here fifteen years ago, but now they come in droves, clogging the highways and the city sidewalks. Sturgeon Bay is full of large new facilities put up by the big national chains— a Wal-Mart, a Burger King, a Hardees, a McDonald's. All the little markets I patronized fifteen years ago are going under. There are a lot of empty storefronts, mostly former groceries, pharmacies, and hardware stores. The nationals and the boutiques are thriving. The original ma-and-pa local stores appear to be doomed.

Fortunately, the fate that has claimed so many of the local businesses on the Door hasn't yet happened to the Hilltop Café. We found that out early on our first full day on the Lakes.

Maybe I should explain first about the Breakfast Game, a pastime I picked up from Rod Badger during the second half of the '83 trip. Put simply, it is a quest for the perfect breakfast. Restaurants are graded on a scale of one to ten through a point system which grants three points for service, three for food, and three for coffee, with a bonus point available for the truly extraordinary. Much of the fun lies in choosing the venue in which to play each morning. A few standard guidelines apply. The restaurant must, of course, be local rather than part of a chain. An older building out in the country or a storefront downtown is usually best. Rod always insisted that the parking lot should be graveled rather than paved; he felt this indicated a concentration on important things, like food, instead of unimportant things, like customers' cars. In *Blue Highways,* William Least Heat Moon—an outstanding Breakfast Gamer—suggested counting the number of calendars displayed on a café's walls, with higher numbers generally indicating better places. Heat Moon, in fact, used this criterion the way a more conventional restaurant critic might use stars: a three-calendar café, a five-calendar café, and so forth. My own favorite guideline has come to be the size of the restaurant's primary identifying sign. Smaller signs seem to be associated with better food—why, I have no idea.

Though I had not yet learned the Breakfast Game when Larry and I passed through Sturgeon Bay in 1983, the best breakfast of that trip—the only one which, in retrospect, could be called a perfect 10—was here, in a small, crowded place called Sawyer's Café. I could find no café by that name in the phone book in 1998, but I did find the

building, and it was still a restaurant. Its sign—a nice, small sign—
identified it as the Hilltop.

Inside it was far less crowded than I remembered, but nothing
else appeared to have changed—not the lunch counter, not the linen-
covered tables in the corners, not even the customers, all of whom
save us seemed to be local people. I ordered the Standard Breakfast (scrambled eggs, breakfast potatoes, toast, orange juice, coffee);
Melody grazed the menu, noting its range and depth with delight.
The food and the service matched memory completely. The coffee,
alas, did not: it was merely very good, not great. In the end we gave
the place a 9.5, and if that isn't quite perfect, it is still very excellent
indeed.

"Was everything OK?" the waitress asked as I paid the bill.

"I've been waiting fifteen years to get back here for breakfast," I
grinned, adding a generous tip to the VISA slip. "I'm pleased to say
it was worth the wait."

She beamed. "Well, we're still here," she said, and I heard the fate
of the other businesses of Door County rustling in the wings. Under
the circumstances—the empty strip mall across the street, the closed
restaurant down the block, the big new Wal-Mart on the far end of
town—"still here" seemed about as positive a statement as could
possibly be made.

And then it was on north, up the Door, under a blue midmorning
sky festooned with little puffy clouds. The countryside was very
beautiful and very rich, with prosperous-looking farms predomina-
ting—mostly the cherry orchards for which Door County is famous,
but also dairy barns and vegetable fields. The highway ran near
the high western side of the peninsula, with occasional views out
toward Green Bay. At intervals it descended to the edge of the water,
and there were tiny towns—Egg Harbor, Fish Creek, Ephraim, Sister
Bay—tucked along bluff-encircled harbors. The towns were full of old
clapboard buildings and steepled churches and green lawns, and they
were far too cute for their own good. Traffic was terrible. In Ephraim
the road ran for nearly a mile through what appeared at first glance to
be a long linear park beside the waterfront; the "park" turned out to
belong in serial fashion to various tourist lodgings across the street,
and all the splendid harbor views—masts bobbing in blue water in
front of the tall limestone cliffs of Eagle Bluff—were guarded by little
fences and "Guests Only" signs. Some kind of festival was going on
in Fish Creek; we lost our way in a forest of booths and tourists and
had to circle several blocks, dodging scores of pedestrians and getting

caught behind slow-moving vehicles looking for nonexistent parking places in the narrow streets, before finding the highway again.

At Sister Bay we escaped down County Road ZZ toward the less-populated east side of the peninsula, away from the crush of tourists, looking for a very special place called the Mink River Estuary. There are only a few natural estuaries left on the Great Lakes these days, and of these few, the Mink may be the very best.

Estuaries are wetlands that occupy the drowned mouths of rivers. They are usually thought of as seaside phenomena, but large lakes form them, too. Their defining characteristic is a chemical gradient. The river flows into the upstream end of the estuary; the ocean or lake, with its very different water chemistry, floods into the bottom end. In between, in the estuary, the waters mix. Organisms can control the chemistry of the water they are exposed to simply by moving a few feet up- or downstream. The multitude of ecological niches this produces, along with the filtering action of wetland plants—which

Mink River Estuary, Door County, Wisconsin (photo by Melody Ashworth)

trap the nutrients coming down the river and mix them with those washing in from the lake or ocean—makes estuarine wetlands among the most productive ecosystems on the planet.

In the oceans, tides flush the estuaries twice daily, increasing the mixing effect and therefore the productivity. In most lakes this piece of the mechanism is absent. In the Great Lakes—astoundingly so to most people—it is present. Tides here are measured in inches, not feet, but that is enough to drive the mixing process. Great Lakes estuaries are freshwater wetlands with tidal flushing, and they are particularly valuable for that unique characteristic.

None of this was understood, of course, when the Lakes were first being settled back in the latter half of the nineteenth century. To white settlers the estuaries seemed merely inconveniently damp pieces of real estate: most of them were rapidly dredged for harbors or filled for building space. Out near the unpeopled tip of the Door, the Mink escaped. With better harbors nearby, it was not a prime candidate for port development. No one was going to build a city upstream, either, because there was no upstream. The odd little Mink is almost all estuary. The river formed during the Chippewa-Stanley stage of Great Lakes development, when Lake levels were much lower, and when the waters came back up they flooded the river valley almost back to its headwaters. Perhaps half a mile of truncated river remains, flowing into two miles of linear wetland, and none of it has ever been developed beyond a few primitive wagon roads and an odd cabin or two in the woods.

Few people knew of the existence of this small jewel when I visited the Door in 1983, and I did not learn about it then. It wasn't until several years later that The Nature Conservancy formed a task force to investigate possible sites for nature preserves in Door County. One of the places they found was the Mink River Estuary. Today the Conservancy holds more than fourteen hundred acres around the Mink—all of the river and its estuary, and most of its small watershed. The old wagon roads have been converted to trails; the few small, weathered board buildings are being allowed to go gracefully back to earth.

The preserve is not well signed, and we were almost past the principal west-side trailhead before we spotted it. We doubled back and pulled the car off the highway. The parking area was merely the end of an old gated road; perhaps two vehicles could have squeezed into it. Fortunately, it was empty. Melody dug out her green cap with the Conservancy emblem on it—she volunteers for the organization in

Oregon—and we headed down the overgrown road into the woods, where we spent about an hour poking around. The woods were lovely: lady slipper orchids, openings with prairie plants, red-osier dogwood with both blooms and berries on it. Sedges lined the estuary, a broad swath of them, waist-high and swaying over the damp earth. We met a young woman and an older one, probably mother and daughter, hiking together: otherwise, the woods seemed devoid of humanity. It was a welcome shift from the madhouse over by Green Bay. Afterward we explored some back roads at the tip of the Door, then came back to Ephraim to take pictures of the harbor among the tourists. It was past 1 P.M. We cut across to Baileys Harbor, on the Lake Michigan side, for a late lunch at a small deli with tables on a lawn, in back, on the water.

Much of what was left of the afternoon we spent at Whitefish Dunes, which were a disappointment. Not the dunes themselves, but the way they were being treated, with the sort of hands-off delicacy that turns people away from environmental protection. Just south of Cave Point—close enough that the two parks share a single access road—Whitefish Dunes State Park protects the largest dune system on the west shore of Lake Michigan. Dunes are common on the east and south shores of the Lake; they are rare on the west. Even among the big dunes near the Indiana/Michigan border, however, these would stand out. They stretch for several miles along the Lake, fronted by flat sand beach; the highest stands nearly a hundred feet tall. All the classic dune features—blowouts, crescents, walking dunes, stabilized dunes—can be found here, often in textbook-perfect examples.

The park existed in rough form in 1983—land acquisition began in 1968, and the place was opened to the public in 1977—but there were as yet no facilities, and it didn't show on most maps. That situation changed in the late 1980s. A visitor center and an immense parking lot were constructed near the park entrance; a trail system was laid out in the interdunal troughs, and a stairway was pushed up the highest dune, named—as most tall dunes along the lakeshore seem to be— Old Baldy. Boardwalks were built across the foredunes to the beach. It was a magnificent beach, and the new access made it easy to get to. For a few years, Whitefish Dunes was one of the most popular state parks in Wisconsin.

That ended in the spring of 1997, when high Lake levels conspired with a series of great storms to wash away most of the beach. The Wisconsin Department of Natural Resources closed what was left to

A deserted beach at Whitefish Dunes State Park, Door County, Wisconsin

all public access except guided tours. They explained this action in a July 16, 1997, press release:

> In the past, the 900-acre state park has won acclaim for its magnificent sand beach. Unfortunately for visitors this year, high water has washed away the beach, forcing swimmers to go elsewhere. Park officials have closed the former beach area until the water level drops and the beach can rebuild.
>
> "By closing the area, we'll be able to protect sensitive vegetation on the dunes' front side, and threatened species such as the dwarf lake iris, dune thistle, and sand reed grass," said Park Manager Allen Miller.
>
> In the meantime, the park naturalist will offer guided tours of the beach at 1:30 P.M. daily.

And that's how things still stood a year later. All non-guided beach access was shut off, and most dune access. Big signs; barricades. The reason continued to be given as "high water," which may have been legitimate in one or two places, but much of what we saw was splendid Lake Michigan beach, totally empty. Dunes and beaches are a constantly changing environment anyway, so keeping human-caused change from occurring is probably futile. I don't want to minimize the possible damage—what would happen if the hordes from the other side of the Door all descended on Whitefish Dunes?—but closing the area entirely isn't the answer, either. If the public can't get access to public land, it becomes much harder to justify calling it "public." In a democracy such as ours, where the people are supposed to be in charge of the government, there are few things more annoying than to come across a fenced-off area signed "Government Land—No Trespassing."

If closing an area were the only way to protect it, this annoyance might be forgivable. In fact, it is not. Visitor education could be improved; smaller parking lots could be built (if overuse is a problem, why provide parking spaces for all those extra people?), and sacrifice areas could be designated where destructive activities were permitted, with care requested for the rest of the park. This last approach would probably be particularly effective here at Whitefish Dunes. Opening one dune for scrambling and one part of the beach for use even during high water would give visitors at least a taste of what they came for while still protecting most of the resource in the strict manner the DNR obviously prefers. It would also let people see, up close and personal, the damage they do to these areas by using them. They wouldn't have to just take the DNR's word for it. With judicious signage, that could be the best visitor education of all.

Anyway, it didn't happen, and as a result most of what we went to Whitefish Dunes for didn't happen either. We did walk the Red Trail for a mile or so, along the first interdunal trough in from the Lake. This is the trail normally used to access the cross-dune boardwalks to the beach, and we passed two of them, blocked off and emblazoned with big "No Access" signs. The afternoon was hot and beginning to go muggy, and there was the big cold Lake rolling up and down its empty beach, inviting but inaccessible, like an ice cream cone in somebody else's hand. I don't know if any of the means I've suggested is the answer, but there has to be a better way than this.

SOUTH OF STURGEON BAY AROUND 4:30
in the afternoon we stopped at a small roadside stand along County
Road S to purchase some fresh Door County cherries as a house
present for Merlin and Cathy McDaniel, old friends from Oregon with
whom we would stay the next two nights. At Algoma we picked up
Highway 42 and came south along Lake Michigan, straining to catch
glimpses of water between the dozens of new subdivisions that seem
to have sprouted since 1983 along the Wisconsin coastline. For a while
south of Manitowac we made an attempt to get even closer to the
Lake, along County Road L, but that proved just as futile for escaping
housing developments as had the state highway, and at Sheboygan
we gave up and hopped on the interstate. We sampled that excellent
Great Lakes tradition, the Friday fish fry, at—of all places—a Wendy's
restaurant just south of Sheboygan, then came on into Mequon, the
Milwaukee suburb where Merlin and Cathy now live, reaching their
house around 8 P.M.

Eastern Wisconsin felt rich. The land we traveled through looked
prosperous, and the Lake seemed recognized as the asset it was
rather than being ignored—a fault which had been preposterously
common in 1983. At least this is one positive conclusion that can be
drawn from the profusion of lakeshore housing. People want a view
of the Lake, and they are clearly willing to pay for it. Agriculture
also seemed ascendant, even near the cities. Houses and barns were
freshly painted; crops looked healthy and green. Could it be that we
are actually gaining ground?

Maybe not. That evening in their family room, Merlin and Cathy
spoke of an election in process to connect the city of Mequon to the
Milwaukee water system. If this connection takes place, Mequon will
be pumping water from Lake Michigan. Drainage patterns along
the west shore of the Lake are very strange, and the edge of the
Great Lakes Basin comes to within a few miles of the shore in many

places; because of this, supplying water to the suburbs of lakeshore cities often means sending it out of the Basin altogether. This was a large issue in 1983, and it remains one today, although it is focused somewhat differently. Then, the fear was of large-diameter pipes sucking the Lakes dry to mine coal in Montana or grow corn in Kansas or cater to the apparently insatiable demands of the water junkies in Phoenix and Tucson and Los Angeles. Today that threat, although still present, appears to have receded considerably. Now the fear is of many small diversions accomplishing together what one large pipe might have accomplished single-handedly.

Two examples will have to suffice here. I learned of the first right before I left home, when a friend in Marquette, Michigan, sent me a newspaper clipping concerning the uproar that a firm in Sault Ste. Marie, Ontario, had caused by announcing plans to ship bottled water from Lake Superior to Asia, where it presumably would appear on grocery-store shelves. The second is more apropos to Mequon, in that it is much closer. A few miles south of Milwaukee, near the Wisconsin/Illinois line, the town of Pleasant Prairie has caused a small tempest in this very large teapot by diverting Lake Michigan water into its city water system. Though a suburb of Kenosha, which fronts the Lake, most of Pleasant Prairie is built along a fork of the Des Plaines River, and the water which enters Pleasant Prairie's taps from Lake Michigan flows out again down the Des Plaines into the Mississippi and the Gulf of Mexico and is thus exported from the Great Lakes Basin as surely as the bottled bits of Lake Superior which the firm in Sault Ste. Marie wanted to send to the far side of the world.

There is also the issue of consumptive use to consider. Even when not sent outside the boundaries of the Great Lakes Basin, water removed from the Lakes may not necessarily be returned. It may water crops, or city lawns, or golf courses; it may be incorporated into manufactured goods which are shipped to other parts of the country. It may evaporate from unsealed reservoirs. Individually, each of these losses is small, but like small diversions, they add up. Basinwide, the Great Lakes system loses nearly fifty million gallons of water to consumptive use each day.

The maps I was carrying did not show clearly which side of the Great Lakes/Mississippi divide the city of Mequon was on. If Lake Michigan water were to flow into Mequon's pipes, would its citizens, too, be exporting water from the Basin? Even if not, how much would they be consuming? The two issues are tied together, and they won't go away.

III
THE SOUTH BASIN

MERLIN MCDANIEL IS EMPLOYED BY THE U.S. Forest Service as a civil engineer. In Oregon he designed water systems and sewerage systems; here in Milwaukee he works out of the Regional Office, doing dam safety inspections. Water issues tend to flow together, if you'll pardon the expression, and so I was not surprised when, upon hearing that I intended to spend much of the next day driving down to Waukegan to look at the harbor, Merlin suggested that he and Cathy might like to come along.

We set out around midmorning in the McDaniels' maroon Mazda sedan. Merlin, who is short, slightly stocky, and bearded, was at the wheel; I was riding shotgun. Melody and Cathy occupied the backseat. We drove down Milwaukee's Lake Drive past what Merlin called "nickel and dime views" of Lake Michigan between large lakefront estates, then doubled back to Interstate 94 and headed for Illinois. By 2 P.M. we had entered Waukegan via Highway 41 and blundered our way down Grand Avenue and East Clayton to Beach Park, at the harbor's edge.

There has been an American settlement on the sandy bluff above the tiny Waukegan River since at least 1829, when a trading post called Little Fort was established there to serve the Potawatomi Indians. In 1849 the local citizens—tiring of the diminutive—changed the name of their growing town to Waukegan, which means "Little Fort" in Potawatomi. No improvement from a strictly denotative standpoint, but it certainly does sound better. The harbor didn't come along for another forty years. There was no natural harbor at Waukegan, just a broad sand flat extending from the base of the bluff roughly half a mile out into Lake Michigan, and it was only after an artificial harbor dug into the sand flat was completed in 1902 that the city—a designated U.S. port of entry for at least twenty years prior to that—had a port of its own to enter.

But the harbor is not really the story here: the story is found in what used to lie beneath the harbor and now mostly lies piled up beside it. In 1983, Waukegan Harbor had the dubious distinction of being among the two or three most heavily polluted waterways in North America. The principal contaminants were PCBs, almost all of them from a single source: an Outboard Marine Corporation plant located between the harbor and the Lake, on what was left of the sand flat. Outboard Marine built Johnson and Evinrude outboard motors, and PCBs had been used in the manufacturing process. This was in the mid-1950s when chemistry was making dreams come true, and the company had no particular reason to be careful with a man-made miracle substance which they had been assured repeatedly was cheap, inert, and nontoxic. Whenever there was a spill in the plant, they simply washed it down the floor drains and into the harbor. Largely insoluble, slightly heavier than water, the stuff sank to the bottom and stayed there. By 1972, when enough doubts had been raised about the material's safety that the use of it was stopped, a large percentage of the harbor's bottom sediment was composed of PCBs.

That was the way things still stood in the mid-1980s, when I researched and wrote *The Late, Great Lakes*. A report on the harbor's contamination published by the International Joint Commission's Water Quality Board in 1982 termed it "grossly contaminated." One bottom sample taken from Slip Three, closest to the Outboard Marine plant's outfall, showed a PCB concentration of 500,000 milligrams per kilogram—a full half of that particular sample. Fish were swimming around with up to 77 parts per million of PCBs in their body tissues, and the water itself was carrying as much as three micrograms per liter, a tiny amount which seems a whole lot greater when you remember that PCBs are supposed to be insoluble in water and therefore shouldn't have been present in the water column at all. Fishing was banned in the harbor, and dredging was suspended until pollution authorities could figure out what to do. There was some sentiment for suspending the dredging permanently and simply allowing the artificially created harbor to revert once more to a sand flat, burying the dirty sediments under twenty feet or more of cleaner ones. Not even its proponents were sure this would really help, however. "What do you do with the sludge?" an IJC official mused to me rhetorically in Windsor, Ontario. "Do you stir it up and remove it? Do you cover it up? If you're going to remove this stuff, where is it safe? What makes it safe? What are the conditions? We just don't know. So is it safer to leave it where it is? I still don't know that, either."

Ultimately—and predictably—the attrition solution to Waukegan's PCB problem was rejected. The infrastructure invested in the harbor simply proved too valuable to abandon. An AOC was declared, a RAP was prepared, and then, under the terms of a consent decree signed in 1989, Outboard Marine and the U.S. Environmental Protection Agency scraped the bottom off of Waukegan Harbor. Slip Three, where the contamination was worst, was enclosed behind a wall of steel sheeting and clay, effectively isolating it from the harbor. Then it was pumped clean. The sludge was treated to remove the PCBs, which were taken off-site for disposal; the remaining material was dumped back into Slip Three, now termed a "containment cell." Spoil from elsewhere in the harbor underwent the same PCB-decontamination process, and then it went into Slip Three, too. When Slip Three filled up, two other containment cells were constructed: one in an area known as the Crescent Ditch, the other in Outboard Marine's parking lot. By February 1992 the job was finished. The bill came to $23 million, nearly ten times what it had cost to dig the harbor into the sandbank in the first place. The "No Fishing" signs were ceremoniously removed from Waukegan's breakwater in February 1992: fish from the harbor still showed PCBs in their tissues, but at no higher concentrations than those caught in the open waters of Lake Michigan.

Some sort of celebration was going on beside the harbor when we finally found it. Part of the parking lot for Beach Park had been roped off in preparation for what looked as though it would soon be a fish fry—though this was Saturday—and a rock band was tuning up on a portable stage. Merlin squeezed the Mazda into a probably illegal spot at one end of a crowded row of vehicles and we walked across the lawn to the water. Sailboat masts did little curtsies and pirouettes as small swells rose and fell beneath them. Beyond, on the far shore of the harbor, a haphazard mass of buildings crowded together on the sand flat, crowned by a tall water tower emblazoned in large letters: JOHNSON OUTBOARDS. A flock of northern shovelers paddled slowly past a sign that read

---

DANGER

Hazardous Bottom
No Swimming

"No one told those guys about the hazardous bottom no swimming," I remarked.

"I don't think those are bottom-feeding ducks," Merlin said. "I don't think it matters unless they get down there and plow through the sludge."

We walked to the end of the breakwater—a long walk, over half a mile out into Lake Michigan. Pleasure-boat slips lined both sides for the first thousand feet; then they fell away. To the left, across a two-hundred-foot-wide channel, the truncated edge of the old Waukegan sand flat paralleled the breakwater. To the right, open water stretched to the southern horizon. Anglers lazed at their rods, some on the channel side, some on the outside. The Lake carried small sails like punctuation marks.

Merlin touched my elbow and pointed south. "Is that Chicago?"

I lifted my binoculars. It was indeed Chicago, a square-edged mountain range of buildings with their feet below the watery horizon,

Waukegan Harbor, Illinois, looking shoreward from near the end of the breakwater

looking flooded, as if global warming had already occurred. Gulls dove through my magnified field of view like feathered cannonballs. Overhead, gray clouds had gathered, pregnant with rain but not quite ready to give birth.

Halfway back to shore Merlin paused, regarding a large brick building on the far side of the channel. Neatly kept and surrounded by green lawns, it looked vaguely out of place against the welter of industrial ugliness behind it. "That's a water treatment plant," he stated flatly.

I wasn't about to argue with a civil engineer whose recent specialty had been water treatment plants. "Maybe it's for the Outboard Marine factory," I said.

"Maybe." He sounded doubtful. "I wonder if they draw the water from the harbor."

"Hard to tell."

"Yes, it is. I doubt it's for the manufacturing facility—it looks too big. They usually just use little box plants."

"Do you suppose it's the city water supply?"

"Out of the harbor?" He smiled. "That would be ironic, wouldn't it?"

Ironic is perhaps not quite the correct word, but I understood his meaning. During the flap over the contamination of Waukegan Harbor, we heard much about the danger of consuming fish caught in the harbor. We heard about danger to the fish themselves, and to the wild animals—primarily birds—that eat them. We heard very little about the potential hazard to the water supply for a city of eighty thousand people.

I have checked a chart of the harbor since. Merlin was right: we were looking at what the chart labels the Waukegan Pumping & Filtration Plant. The intake is not clearly indicated on the chart, but it appears to be out in the open Lake, not in the harbor. Still, the irony—if that is the right word—is there. Twenty-four million people drank the Great Lakes in 1983; the numbers are up to roughly 27 million today. I don't want to belittle the problem of contaminated harbor sludge, but what of the waters themselves? What are we to make of the fact that many of the confined disposal facilities created for toxic materials dredged from the bottoms of the Lakes are located, not on land, but—like Waukegan Harbor's Slip Three—directly in the Lakes' waters?

Good old *Homo sap.* We are indeed a clever race. It's too bad we couldn't be smart, as well.

The rain began shortly after we returned to the McDaniels' car. It was not a heavy rain, just a steady drumming on the windshield that made us glad we were no longer out on the breakwater. The car glided up Highway 32 along the lakeshore, which—like that just south of the Door—seemed mostly dedicated to housing developments. We passed the Wisconsin line. I was about to remind Merlin that I wanted to at least drive through Pleasant Prairie when I noticed we were there. A small sign beside the road announced the Pleasant Prairie city limits. I glanced to the right. The highway was riding the top of a low, flat ridge. Off to the east, beneath heavy clouds, glinted the great silvery expanse of Lake Michigan.

"We must be on the divide," Melody remarked from the backseat.

Right enough, and that is Pleasant Prairie's dilemma. It is easy enough from a hydrologic standpoint to define the edge of the Great Lakes Basin; it is far harder to identify the same divide in the human mind. Where is the psychological edge—the ridge that separates those who consider themselves people of the Lake from those who do not? Given the immense variability of the human psyche, this boundary between who is a Lake person and who is not is bound to be fuzzy; but certainly if you can see the Lake from your town you should qualify. And if you are a person of the Lake, should you not be able to drink it?

It is perfectly, straightforwardly, irrefutably reasonable to state that water should not be exported from the Great Lakes watershed. But here at the southwestern corner of Lake Michigan's South Basin, the Great Lakes watershed does not have a perfect, straightforward, even vaguely reasonable boundary.

Back at Merlin and Cathy's we harvested corn from their garden—on land which, I was assured, was tributary to the Milwaukee River and thus safely within the Great Lakes Basin—and grilled it on their patio under clearing skies, along with chicken and foil-wrapped potatoes. Then we talked until past midnight. Merlin told us a little of his work. The Eastern Region of the Forest Service is an agglutination of the old Eastern and North Central States Regions: it is headquartered in Milwaukee but encompasses the entire northeastern quarter of the country, from Maine to Minnesota and from Maryland to Missouri—twenty states in all, plus part of a twenty-first (Virginia). Within the Region, and especially within the Great Lakes states, individual forests were also undergoing agglutination. That summer of 1998, the Chequamegon and Nicolet National Forests in Wisconsin—administratively linked for years—would be formally merged. The Huron and Manistee in Michigan's Lower Peninsula would soon follow. There was talk of combining the Ottawa and Hiawatha in the Upper Peninsula, and the Chippewa and Superior in Minnesota. The mantra, each time, was "government efficiency." Reduced funding requires streamlined operation. Merging forests lowers administrative overhead, allowing the same services to be offered on a smaller budget.

Like the out-of-basin diversions argument, the move to combine National Forests is perfectly straightforward, logical, and reasonable on the surface but runs quickly aground on the shoals of the human sense of place. Before it can be managed properly, land needs to be understood with the bones as well as the mind. A forester in Milwaukee may gather excellent information about a forest near Marquette, and may issue management decisions with perfectly good intentions; but the forest still would be managed better if the forester actually lived in Marquette. Good information and good intentions do not automatically make good decisions. There are always Pleasant Prairies in the world, straddling management zones, and they create issues that cannot properly be resolved without a direct feel for the milieu in which the resolution must play out.

The movement to reduce the size of government thus runs up against a paradox. To shrink government in manpower and money requires enlarging it in terms of the physical units being governed, and this unavoidably increases the distance between the government and its citizens. If the federal presence is to be truly small in scope, with local people making local decisions, then staffing and budgets need to be *increased,* not shredded. If we really want to get control back in the hands of the people, that is the direction we will have to go.

MADISON, WISCONSIN, IS A GREAT LAKE
city, but it is not a Great Lakes city. The capital of the Badger State
is built on a narrow isthmus between Lakes Mendota and Monona,
and the little Yahara River—on which these two medium-sized lakes
and their smaller neighbors are strung like large watery beads—
flows south to the Rock River in northern Illinois and thence to the
Mississippi and the Gulf of Mexico. Geographical minutiae aside,
however, it was still my focus on the Great Lakes that brought me
here. Wisconsin has long coastlines on both Lake Superior and Lake
Michigan, and Madison is full of government workers whose jobs
revolve to a very large degree around Great Lakes issues. It is also
among the most livable cities in the country, and this—plus easy
access to state government—has made it a magnet for environmental
activists, who revel in the fact that they can enjoy the amenities
Madison offers and still be closely connected to one of the region's
major political ganglia.

In 1983 I had interviewed two people in Madison, Sierra Club Midwest representative Jane Elder and Wisconsin Department of Natural
Resources water specialist DuWayne Gebken. Both interviews had
been fruitful. Jane had proved to possess a deeply analytical mind
coupled with extremely broad knowledge of Great Lakes issues;
DuWayne had turned out to be that rarity, a management-level natural resources official who still thinks like a field biologist. Each had
been passionately committed to the defense of the Lakes. It seemed
prudent to go back and see what they could tell me today.

We drove from Mequon to Madison on Sunday afternoon and
immediately proceeded to get lost. Not seriously lost, just enough
to jar the composure a bit. After a half hour or so of confusion
we finally found the motel that held our reservation—barely three
blocks from the freeway exit where we had begun our mini-odyssey
through Madison's street grid. We checked in quickly and left without

unpacking the car. We had a rendezvous to make with Jane: not at her home, nor at her office near the capitol, but at campsite number six in Blue Mound State Park, thirty miles out of Madison to the southwest.

The original idea, actually, had been to meet at a brewpub near the university. Jane, who no longer works for the Sierra Club and therefore feels somewhat out of the loop, had offered in one of our early e-mail exchanges to bring along a group of currently active Madison-area environmentalists whose views she thought I should hear. It was one of those excellent plans which circumstances conspire to doom before they can get off the ground. The day before I left home, Jane called to cancel the meeting. She sounded very apologetic. The evening Melody and I would be in Madison, she had belatedly realized, coincided with the night she and a friend (also named Jane) had set several months before to introduce their preschool sons to the joys of camping. The campsite reservation couldn't be changed now. She hoped I would understand.

I did; but I also didn't want to miss the conversation with her. Would it help if I shifted my motel reservations a day earlier?

"That might be a little hard to do in this town," she responded, somewhat doubtfully. "Maybe I could cut the camping short and come in for lunch on Monday. Or you could come out to the park," she added, brightening up. "It's only half an hour out of town."

And that's how we ended up at Blue Mound, which is a displaced outcrop of Door County dolomite humping up out of the rich, flat farmland of the Driftless Area, the only part of Wisconsin the glaciers didn't tear to shreds. We bought locally made cheese and locally grown carrots at a tiny store at the base of the mound and headed into the park, where Jane—a tall, slim blond in a white T-shirt with a bright graphic of cats splashed across it—greeted us warmly. She introduced us to her husband, Environmental Defense Fund attorney Bill Davis, and to three-year-old Colin.

"I've been trying to remember when we last saw each other," I said. "I think it was at Love Canal."

She smiled at the recollection. "Bill and I participated in a rather bizarre bus ride around Niagara Falls," she explained to her husband and my wife, "back in—was it '86?"

"'87. The GLU convention." Great Lakes United's annual meeting had been held in Niagara Falls, Ontario, that year—the only year I have so far been able to attend.

"Right. What made it so bizarre was that it was a tour of toxic waste dumps. Starting at Love Canal, which they were in the middle of cleaning up."

"And ending at Model City, where they were dumping the junk from Love Canal."

The other Jane—I never learned her last name—arrived with five-year-old Michael in tow, and we sat down to a shared supper around the campsite table. Colin and Michael burned their toasted marshmallows and had to be convinced they were still edible.

"So what are you doing these days?" I asked. "On the phone, you said something about a biological diversity project."

She nodded. "Let's see," she said, reflectively. "Short history. I left the Sierra Club in the summer of ninety—get my years right—five. Colin was six months old. And I took a job with something called the Consultative Group on Biological Diversity. They had done a series of focus groups around the United States, trying to figure out why American response was so lukewarm on the Global Biodiversity Treaty. And based on that, they decided that it was time to look at a comprehensive new strategy for trying to get the public to understand why biodiversity is so important, and why there's a global crisis. How do you help the public grasp a concept like biodiversity? Is it media? Is it education? There are things we're recommending that we're hoping will be adopted or nurtured by environmental groups."

"Is this a national effort?"

"Yes. Primarily U.S., a little bit of Canada. A couple of Canadian funders are in the group. We're not spending a lot of time on global biodiversity—it's in the back of the agenda. The real interest is in how to get the American public to sit up and pay attention."

I asked if she was familiar with the biodiversity report that had been done by the Wisconsin Department of Natural Resources, which I was carrying around with me in the car. DuWayne Gebken was proud of that report. Jane hadn't seen it, and she didn't appear to have much interest in seeing it now. "A lot of people try to send us the policy stuff, and the inventories, and that's not the end of the thing we're working on," she explained. "We're working on what the public gets. A colleague of mine from The Nature Conservancy was reading our final report, and he said, 'What about prairies and deserts? They're under-represented in this report.' And I said, the American public doesn't yet *care* about prairies and deserts the way they care about forests. And yes, we want them to care about these things, but they're not there yet." She laughed. "So part of our job is to translate where the public is—what people will respond to, and what they want to talk about—to folks who are in his position. It's been a fascinating transition from some of the front-lines work in the Sierra Club."

"What made you leave the Club?" I asked her.

"You really want to know?"

"Sure."

"Downsizing. Three days before I went into labor my boss called me—"

"Oh, jeez—"

"—and said, 'As you know, by the fifteenth of November we have to make all these decisions, and we don't have the money to keep you, and I also can't imagine tossing you out.' So what they offered me was 56 or 57 percent time when I came back from maternity leave. With no support staff. We tried it for about six weeks, and I said, 'This stinks'—" We all laughed. "And that's why I left. I thought, after all these years, if this is what I have to do to stay . . ." She sighed. "I didn't want to leave. I was invested in the ecoregions program, I was having a lot of fun—and now I get to watch other people do it differently."

"And wrong, of course."

"Well—differently. You know, I'm in a different place, and they're in a different time, and the politics are different. I still share office space with the Sierra Club. There were three little spaces, and we needed three. I knew the people, I knew the Xerox machine—" She smiled. "I knew the parking spaces, and I said, Hey, what do you think? So they're my landlords now."

"So you're still in the same office."

"Yes. And it's a mostly pleasant relationship, because I get a sense of what's going on, but I don't have to get involved in it. And it's hard sometimes, to watch decisions that I think are not the right decisions—"

" 'Mostly' is the operative word."

"Mostly."

There was a short silence. Beyond the gathering darkness at the table the campfire clicked and flared.

"So let's talk about the Lakes a bit," I prodded. "We were reminiscing earlier about Love Canal. That was one of the things that was hot fifteen years ago, and nobody talks about it much anymore. What else do you see that has changed?"

She thought for a moment. "I think the biggest change is what I would call the Republican shift, or the Republicanization of the Lakes," she said at last. "We were active at a time when there was tremendous energy, although sometimes you don't recognize it when you've got it. We now have almost exclusively a Republican, pro-business government. You have a governor like Michigan's John Engler, whose attitude toward fish advisories is, if people don't know about them then the fish are clean. The momentum has shifted, and I don't think we've ever quite recaptured it. There's activism, there are still people doing things, but it's different. I think the political shift has been the hardest thing for us long-timers to acknowledge. A lot of us are still in denial.

"Look at the International Joint Commission. There was a body that was considered to be learned, independent—and *useful*. Their scientific reports were groundbreaking for the Great Lakes. They drove the agenda. But in the last ten years the IJC has become a very different animal. It's extremely politicized, in terms of the appointees. Industry has discovered it. So the instruments have changed.

"Part of what has been interesting is that much of what has been taking place recently on the environmental front happened first here in the Great Lakes and then went elsewhere. All the attention on endocrine disrupters in the last four to five years—we were doing en-docrine disrupters before anyone called them endocrine disrupters. The same toxic chemicals that people were paying attention to in the Great Lakes are now the ones at the heart of the controversy. So there's a gratifying sense of being part of riding that crest, and pushing that agenda. I don't know what the current agenda is. There's a lot of attention on mercury. People get all excited about the really scary endocrine disrupter stuff, but the carcinogens are still there every day. We still have issues of water quality. Closed beaches. These things seem mundane, but—I remember talking to a woman at a conference a couple of months ago, from Racine, and she said, 'What a crime'—this was summer in a city on the Lake—'When I was child growing up,' she said, 'summer *was* the beach in Racine. And now it's—is it safe today? Can I take the kids today?' Sewage issues."

"We drove down the Fox River a couple of days ago," I said, "and saw people water skiing, and we wondered if they knew exactly what they were doing."

"Well, on the Fox, so much of the mess is in the bottom," Jane pointed out. "And people have never really understood contami-nated sediments. You don't see it—you don't interact with it—and if you do see it, it's just some big scoop of glop coming up. It's out of sight, out of mind for most people. Lake Erie is curious that way, too, the water quality is so much better than it was—"

"That's the zebra mussel."

She smiled. "Aren't they amazing? It's like, 'Stop filtering the water! *Slow down!*' But there have been weird things like that going on. We're still getting these, you know, bizarre effects from the exotic aquarium. But I don't know anybody, ten years later, who loves the Lakes less.

"What we've found in the research I'm currently involved in is interesting. Environmentally, we had assumed there was a sense of the Great Lakes region—that people thought of the Great Lakes as a place. But most of the research shows that people don't identify as being from the Great Lakes region. They don't identify themselves with the Lakes. We were trying to build a movement around a perception that was not grounded in reality. It's had a lot of people rethinking things. I'm a romantic, and I can't imagine another way of taking on the issues. But from the organizations to the foundations, a lot of people are asking—does it make sense to organize around the Great Lakes, as a Great Lakes environmental movement?"

"Why do you think that is?"

"Well, politically we're organized by state or province, not by large ecosystem. There's issues of the U.S. and Canada all the time trying to figure out cross-border policies. And I think, because of the culture, the population doesn't identify with the Lakes as *their* system, *their* Great Lakes. You're always up against bringing the public along. So the question is, where *can* you find them and what *do* they care about?"

"Part of it is just a matter of scale."

"Part of it's scale. People don't think in that scale. We saw that in some Mississippi River research in the Twin Cities. People were asked what their most common encounter with the river was, and they said 'Driving over it.' And there they were, near the headwaters of one of the world's great river systems. We just don't interact with systems as systems. So—I'm not sure what the next step is, how you build that consciousness."

"I've been back here several times since I did the book," I said. "You may recall that in the book I made a comment about apathy around the Lakes. My sense has been that it's fading some. Lakefronts are being developed facing toward the Lake instead of away from it—people are talking about the Lakes. They seem to be more aware of the issues."

"Oh, I think, clearly, ten or fifteen years of activism and agitating pays off," Jane agreed. "And even though Governor Engler doesn't want you to know about the fish advisories, people who fish know. Not all of them. But the consciousness is out there. There's an awareness that the Lakes are in trouble, on a certain level. And we're discovering that they're an asset—that people *like* water, and open space, and views. There's even talk about bringing cruise ships into the Great Lakes again. But it's really hard for me to judge, because I'm in a very—you know—biased set of friends, and such. When it comes to the Great Lakes, my blinders are pretty big.

"It's interesting, though, the development issues around the lakes. People are sensing the loss of shoreline, the loss of open space. We're losing the North Woods—we're losing the natural shorelines, the wild beaches and the long walks. There's a sense that we have to *do* something. Sprawl is very hot in the region. It's funny, because land use was such a big thing twenty or twenty-five years ago, and now it's back again. There's a recognition that we're losing the rural landscape. And certainly the public gets it better than the politicians. Why they keep *electing* these clowns is beyond me." She laughed. "It's the economy. I guess that's it. It's funny, because the environment is always high, in the upper Great Lakes states, in terms of voter consciousness. But it hasn't been driving elections in the region. Maybe a handful here and there. But in gubernatorial races, in particular—we continue to elect people whose passion is Great Lakes industry."

I nodded. "And funding for the agencies continues to go down. Take the RAPs. We were all looking at them as something that might actually do something, ten years ago. And today most of the agencies involved are dropping funding for them, and the work has not been done. So how do you read that?"

"Well, they were never mandatory." She folded her arms across the cats. "The Remedial Action Plans were part of the revisions of the Water Quality Agreement—they came out of that process. And a number of us, when we were trying to codify the Agreement into domestic law, said, 'You're crazy, we'll never get it through.' Well, we sort of did and we sort of didn't. But without the force of law behind it, you can't sue. There's no hammer."

"Some of them were actually completed, though," I pointed out.

"Yeah."

"I'm thinking particularly of Collingwood. It was actually taken off the list."

"And they're crowing about that." She laughed again. "Most of the RAP work—it was not my focus, and so I hesitate to speak with any sort of authoritative voice on it—but a large part of it was process. A lot of bureaucracy—God knows how many meetings—"

"And nothing, basically," I said.

Jane nodded. "I used to challenge my buddy Tom Martin, who worked for Governor Blanchard in Michigan. And after we'd go to these meetings, and we'd *go* to these meetings, I'd say to Tom, OK, how many tons, and how many pounds, got cleaned up today?"

"So let's talk briefly about the Detroit River," I said.

"It's still there." Jane smiled.

"Yeah," I said. "Yeah, but they've dredged it."

"I don't know what they've done with the dredge spoils," she said. "Do you?"

"No. I was hoping you did."

"Nope. You've got me. You know—my rusty, dusty Great Lakes days." She sighed. "Let's just say I'm glad I live upstream. It's been interesting, though. When I talk to foundations now, as part of this biodiversity project, I ask people where their drinking water comes from, and it's amazing how few of them know."

"OK," I said, "let's talk about drinking water for a minute, and let's talk about diversions. It was always a hot issue."

"Yeah."

"It's still a hot issue."

"The Canadians." She laughed again.

"The ones that are selling the water to Japan?"

"Yes. I understand that's been derailed, I'm happy to say."

"I hadn't heard that part yet," I told her, "but I'm sure I'll pick it up from several other people. Clasping their hands in glee—"

"But it keeps coming back. It's like the Corps of Engineers. Any time you kill a Corps of Engineers project, within three years it has a new name, and a new general, and it's back. And I think the Lakes are going to increasingly be a temptation to a world that has only so much fresh water. The idea of shipping Lake Superior water halfway around the globe—it's just absurd! And yet, people are willing to pay. Look at how much we pay for bottled water. Lake Superior's going to continue to look really good to people. And I think that neither the governors nor the congressional delegation is really anticipating the whole global freshwater shortage. In our lifetimes we're going to see the Great Lakes threatened again and again and again, as a source of good water."

"Let's back up from the big schemes," I suggested. "Let's look at the little ones. We drove through Pleasant Prairie the other day, and as you're at the sign on the highway that says 'Pleasant Prairie,' you can look over and see Lake Michigan. And yet most of the town lies outside of the basin, because it's in the Des Plaines watershed."

"Well, it's the same with Chicago and Milwaukee."

"Right. So how do you tell these people, who can see the Lake—"

"That they can't drink it?" She propped her elbows on the table and interlaced her fingers. "That's a huge issue. Part of it is not understanding that the Lakes are busy being the Lakes. Without that background, people go—'But look at it! It's *right there!*' And it *may* make sense to do water supply to those cities. But then, how far out do you go?"

"Right."

"And how many exceptions do you make? And it honestly just amazes me that people think they're big bathtubs."

"Part of the question also revolves around what's coming back in through the sewers," I said. "We were staying with a friend in Milwaukee who's a civil engineer, and a point he made was that water taken out of the Lakes will go right back in if the sewers dump that direction. He also wondered about where else Pleasant Prairie's water would be coming from. If it's coming from groundwater, the cone of depression may well go down to the Lake anyway." A cone of depression is the hole a well makes in the water table as you draw it down; the deeper the drawdown, the broader the top of the cone. As Merlin had pointed out, these cones are not bound by surface topography, and they often reach out several miles from their wells— more than enough for a well in the Des Plaines watershed to pull groundwater from the Lake Michigan side.

Jane nodded. "They're tricky issues," she agreed. "The impressive thing about where we are with environmental decisions in the nineties is that the simple decisions have all been made. The easy laws have all been passed. And now they come down to—if you do this, this might happen, and if you do that, that might happen, and there are all these shades of gray. And it doesn't address the larger issue of how many people the Great Lakes Basin can support. People in this region do not think of water conservation. We are a water-rich region. Californians are very conscious of that kind of stuff, but in the Great Lakes region people just don't get it. We just turn on the faucet. We joke about interbasin transfer every time you have lunch in Madison and go pee in Milwaukee—" We all laughed again. "And how much are we exporting in Great Lakes beer? That's another question."

"Yeah, that was a question that came up last night, too, after a couple of them."

Jane was serious again. "I think . . ." She paused. "The challenge is that it's a geological system. People can't think at that scale, and we don't think five years down the road. The Great Lakes decisions that we're making are happening on large scales that politically we're not designed to deal with, and socially we don't want to think about, or interact with. In many ways, the Great Lakes are a microcosm for other big global environmental issues. We don't have the political systems, or the cultural mind-set, to deal with these things. And that's where it's sort of scary as we look at an issue like global warming or biodiversity loss. If you can't save the world's largest freshwater system—which is *right there,* in 27 million people's backyard—how do you get people to address two-hundred-year-long issues? It's tough. So—education, and politics, and democratic institutions—anytime we scratch something in the Great Lakes we end up with these large cosmic issues."

"We're getting into the issue of sustainability here," I pointed out. "Do you know of studies that have been done on sustainability levels of the Great Lakes—that is, the population level that can be supported here in the Basin?"

Jane turned to her husband. "Bill, that's your thing, really," she said.

"Nobody's done that work," he said gloomily. "It's one of the projects that might be really useful. But most people don't seem to care."

"But in Gary and northwest Indiana," Jane said, "the steel industry's changing. Absolutely changing. There's a shift. It's hard to pin down. But some of the big industries, like steel, are on the edge already. There's a change in the wind. It's hard to know where it's going, and it's hard to say whether it'll be enough, but it's been fascinating to watch. There's an undercurrent in the American population now that's beginning to question consumerism, and how much stuff we have, and materialism. And those are the sorts of things that give you hope. Only when we change those basic lifestyle patterns, and those consumption patterns, is sustainability going to arrive. To the extent that people begin to recognize these things on their own, there's hope. We don't know if the planet has time for people to catch up. But we've never known that." She paused, reflecting. "When we were up at Isle Royale four summers ago we saw a fledgling pair of bald eagles. They were hatched on the island, eating island fish. We sat up on the ridge and watched this young pair work, and—one, they were beautiful, but two, what they symbolized was really exciting. Now both of that pair have mates and babies. It's really good news."

"Is this OK?" Colin interrupted, brandishing a black marshmallow. "It's burned up."

"Oh, that looks pretty good," his mother exclaimed.

"But it's pretty burned," said Colin.

"You're supposed to burn marshmallows," I told him. "It's one of the rules."

"If you want to just burn that one up right into a crisp and start over fresh, you can do that, too," Jane said reassuringly.

"Want to eat it?" said the other Jane. "Well, take it off with your fingers. I think—whoops!" Colin bobbled the marshmallow and caught it again.

"Or," said Colin's mother, grinning, "you can just eat it right off the ground."

"I think it's paper towel time," I said.

"Um!" said Colin, with his mouth full. "This is good!"

"By the way," said the other Jane, "I hope you aren't terribly mad at me. He had a wonderful time, that's all."

"What's he covered with?"

"Just sand. It's everywhere, though. And since you're sleeping with him—"

"Well, he gets the old sleeping bag, that's all right. Oh, God, Michael—" Michael appeared to have rubbed charcoal over most of his face, and his chin was sticky with marshmallow.

"He's having a great time," said Michael's mother.

DuWAYNE GEBKEN IS A GREAT SHAMBL-
ing bear of a man who overflows a Dilbert-style cubicle on the sixth
floor of a Wisconsin state government building known by the cryptic
and unlovely name of GEF-II. He was away from his desk when
we arrived early on the morning after Blue Mound, but we didn't
have long to wait: soon he came striding down the aisle between the
cubicles, a thick sheaf of papers clutched in one huge paw.

"DuWayne Gebken," he said, unnecessarily, as my hand disappeared into his.

"You haven't changed a hair," I assured him.

He grinned. "Maybe not the hair. The waist and the spectacles are
both a bit thicker. Come on in and sit down."

I introduced my wife to him. Melody is a small woman. It was
a little like introducing a reed to a redwood. DuWayne disappeared
briefly to locate an extra chair, and we all settled into his office. And
those who are uncomfortable with the exploration of ideas might
want to skip ahead here, because a conversation with DuWayne
Gebken always ends up stretching your mind into new and unex-
pected paths.

"Last time we met," I began, "I think we were talking mostly about
winter navigation. But that appears to be pretty much a non-issue
these days."

DuWayne nodded. "Winter navigation ran afoul of interest rates,"
he said. "The profitability was really quite marginal." Winter navigation—the extension of the Great Lakes shipping season through
the winter months—has been pushed by the U.S. Army Corps of
Engineers for many years. But they have found themselves fighting
stiff opposition, not only from the environmental community—which
they had expected—but from the shipping industry itself. Quite apart
from the safety issues—which are significant—there are questions of
competitive balance. The industry has traditionally used the winter

shutdown to do maintenance. With everyone idle at the same time, this makes sense. With traffic moving year-round, it would be iffier: there would always be a concern that someone else could move cargo that was rightfully yours while you were in dry dock. So the shipping industry joined forces with environmentalists, who were concerned about habitat and wildlife survival and endangered species, and lakeshore property owners, who were concerned about damage to docks and beach walls, and that unlikely alliance prevailed: the Corps withdrew their proposal for further study. Since 1983, Lake shipping as a whole has suffered a downturn, raising new questions about the need for season extensions. The issue hasn't gone away—as Jane Elder pointed out, Corps projects never really go away—but its profile is so low these days that it can barely be seen.

"The Corps was trying to apply adaptive ecosystem management to winter navigation," DuWayne said. "They were saying 'Let's do this and see what happens,' and we said '*Wait* a minute!' The stakes were too high for that. Everybody agreed we didn't know enough to predict the results, and we didn't think that was a real good way to find out."

"I remember you complaining that we knew nothing about the winter ecosystem of the Lakes," I said. "You called it a 'virtual black hole.'"

"It's still a black hole. We haven't had the research money to look at it, and I doubt we would know very much if we had. Look at all they've thrown into Chesapeake Bay—there's still so much they don't know. That says to you that money isn't the solution." He gestured. "Size alone in large systems can make a big difference. We just don't have a handle on that. We're in the process of beginning to develop indicators, and we're looking for fish that are indices of biotic integrity. One of them I've seen proposed is the lake trout, for oligotrophic systems. But the lake trout have largely disappeared. So are there better things, like the sculpin?"

"So many things have messed around with the lake trout," I said. "But I've heard recently that there are self-sustaining populations again."

"There always have been, in Lake Superior."

"What about yellow perch? I hear their populations are collapsing in Lake Michigan."

"I don't know that anybody knows yet what is going on," said DuWayne, "except that there aren't a lot of female perch around. Apparently there is some reproduction. They do get to the larval stage. What happens to them after that we don't know. There's some thought that it could be predation from the white perch, or from something else. There's just starting to be investigations. But we could have a fairly significant problem, because the Friday night fish fry in Wisconsin was a tradition that was based upon yellow perch." He looked thoughtful. "Some of us have bandied around kind of a notion of—this is a complicated situation to explain, but have you heard of chaos theory?"

"I've done a little bit of work on that," I told him. "The mathematical equations that represent population dynamics turn out to be strange attractors." Strange attractors are mathematical objects which were discovered in the mid-1960s. Their equations are characterized by what mathematicians refer to as "nonlinearity," which means that—unlike simple formulas such as those for calculating the area of a circle or determining the speed of a falling object—changing the size of the variables in a strange attractor's equation will not always lead to proportional changes in the result. The attractor representing population growth is known as a "logistical hump," and the manner in which it behaves has turned out to be governed by the variable in its equation that represents what biologists call "recruitment rate"—the number of young that reach breeding age. At low levels of recruitment, the population graphs out as a single rising line. At higher levels it becomes bimodal, with some numbers generating strong population growth and other, equally likely numbers generating population crashes. At higher levels yet, the graph goes chaotic. At these elevated levels of recruitment, population size is totally unpredictable from year to year. One year there may be record numbers of the species present; the next year it may be on its way to extinction. It is suspected today that chaos, not overhunting, is what really killed off the passenger pigeon.

DuWayne leaned forward. "And, see," he explained eagerly, "given a point in time in a population cycle with natural fluctuations, if you try to maintain a sustainable harvest what you may do is drive it into a chaotic state. If you start at the rise in the population you may set your targets too optimistically, and drive it into a state that is more unstable, from which it then collapses. All this is, is a notion so far, but it's one that is starting to show up increasingly in the literature. First of all, it was modeled that this could happen—this is, for biologists, a relatively new business within the last couple of years, population modeling—and then it showed up in an experiment in the laboratory with flour beetles. Then it was done in the Chihuahua Desert with rodents. And the latest I've seen was some work that was done on Dungeness crabs on the West Coast. And you're dealing with a fairly wide variety of animals now—an insect, a crustacean, and a rodent—so it makes you wonder whether some of the other things that go haywire might not be due to that, but we simply were not able to recognize it."

"Try to get the politicians to grasp that," muttered Melody.

DuWayne nodded. "It's the kind of thing that makes you wonder whether the whole notion of management, and what we call sustainability, is possible," he said. "And if it's possible, how do we set about doing it? As you know, with chaos it's the initial conditions that make the difference. Where do you start? If you start at some point, and the point that you start at is extremely optimistic, and you drive it into a new state—how do you get off of that?"

"You can't manage it," I marveled, "because you'll destroy it."

Melody laughed. "*That* they've grasped," she said. "But they've really hardly begun to understand Newtonian cause-and-effect-type stuff. And if you try to tell them, 'Well, we have some effects, but they're unpredictable, and until we've seen what they're doing we won't know how to take the next step, and we may have to remodel the whole thing—' "

"There's another thing that's happening somewhat sequentially with this," said DuWayne, "that might also lead us down a path that we're not aware of, and that is how we test hypotheses. We tend to test a hypothesis against a null hypothesis. But does nature work that way? No, it *doesn't* work that way. Nature tests multiple hypotheses at the same time." He laughed. "So that's some of the things we are now starting to deal with. How do we go about doing science in the face of a great deal of uncertainty? New tools and new techniques are being developed all the time which require even more statistical powers of analysis. It's the kind of thing that's caused us to wonder—given all of these things we are seeing as possible problems—even though we have used the best science, might we still not have done the right thing?"

"Well, that's always the problem with management," I pointed out, "because you're always working with inadequate data. And we constantly run into the problem of people who thought they were doing the right thing who find out later that it was precisely the opposite of what they should have been doing."

"Yeah. That's the case many times, and it's one of the things about adaptive ecosystem management that we're beginning to revisit. The notion of a 'project' as an experiment. There's something that was called for in NEPA long ago, but it's been neglected and never followed through on, and that is the notion of monitoring. It's something that, if done at all, has been done superficially, or very poorly. So we're back almost to where we were fifteen years ago, looking at things and saying, 'Well, there might have been something about this that we should have been paying more attention to. But we cast it off to the side.' " He sighed. "We have a lot less certainty than we thought we had. And things are a lot more—I don't want to say 'perilous,' because it's not as dire as perilous. It's more like we just don't know what is coming out of this, or what *might* come out of it. We're going to need a lot of help from the public. We need to have a lot more minds than ours involved in this, and a lot more people involved. Science is searching for help. The science of management of natural resources is looking for help from people who might look at the world in a different way, and see it in a different way than we do. And it's easy for people to say, 'Whoa—you guys are supposed to know this stuff' "—he laughed—"but as we look out, and we see where things are going, it's just not as cut and dried as we thought it was."

"I guess—" Melody paused and started over. "Two things," she said. "One is that management scientists are having to acknowledge that the world is less certain than they had hoped. And the other is that it's perilous to try to convey that to the population at large and get their help, because it's scary."

DuWayne nodded. "It is," he agreed, "in that we have to admit that we don't have all the answers. And if you look at the notion of chaos, what we're saying is, 'Not only do we not have all the answers, but we don't even know where we started from, or what that means.' I think it deals a large hand of uncertainty for the future. It's kind of an unsettling feeling, like we're heading into a maze. You can learn a maze by accident, and by trial and error. That's probably what we're going to do. But hopefully as we get into those blind ends we will get to know a little more about it, and start to develop some rationale for dealing with it."

When we came out of GEF-II, the sky, which had been misting a light rain onto Madison as we arrived, had cleared to a mottled gray-blue that indicated the start of a hot and humid afternoon. We drove to Chicago by back roads that seemed curiously devoid of traffic considering their proximity to the nation's third-largest city; but then, traffic numbers are probably driven by strange attractors, too. By nightfall we had settled down with relatives in Naperville,

Illinois, on the southwest edge of the Chicago metropolitan complex. We were barely thirty miles from Lake Michigan, and there was nothing between us and the shore but flat cityscape, but geographically speaking we were on the Mississippi. That is where the little Du Page River, which flows through Naperville, ends up.

"One of the dilemmas you have," DuWayne had remarked at another point in the conversation, echoing Jane Elder's concern, "is how people perceive where they're at—how they perceive their sense of place. That's something you have to know to manage resources. What do people look at to define themselves?"

Here in Naperville there is no question that they look to the Lake, even as the rain that falls on them flows away toward the Gulf of Mexico. I don't know the best way to develop the "sense of the Great Lakes region" whose absence Jane had complained of, but I suspect that denying a Great Lakes connection to people who already think they have one is not a particularly auspicious start. To care *for* something, it is first necessary to care *about* it. It may well turn out that allowing the little, close-in diversions to proceed is the best way to prevent the big, distant ones from taking place.

IT IS QUITE APPROPRIATE TO BEGIN A DE-
scription of Chicago by pondering the distinction between big and
little Great Lakes diversions, because Chicago is the home of the
biggest diversion of them all. This diversion is also among the oldest,
having begun in 1900. I refer here to the reversal of the Chicago River,
an engineering feat of which—as the old sarcasm goes—there is rather
less there than meets the eye.

Twelve thousand years ago, the land that is now Chicago was
occupied by one edge of an immense lobe of ice that had pushed
its way southward up what had previously been a broad, shallow
river valley. The ice had roughly the same outline as Lake Michigan
does today. It retreated by stages. Each stage left behind a moraine—
a long linear pile of rubble heaped up by the ice along its leading
edge. It is these moraines which are responsible for the odd nature
of today's South Basin drainages. They lie parallel to the lakeshore
only a short distance inland, and they run for miles. Rivers occupy
the valleys between them. Where the inner moraines are breached,
the rivers run into the Lake. Where the outer moraines are breached,
the rivers run away, down the general course of the land toward the
Mississippi and the Gulf of Mexico.

There is one great break in this pattern—one place where the
moraines never developed. A continental glacier creates a great deal
of meltwater, and that water must go someplace. In the case of the
Lake Michigan ice lobe, where it went was south, across the glacier's
outwash plain and into the ancestral Des Plaines River. The gap
where this torrent of icemelt passed through the moraines is today
the site of Chicago, and what is left of the old meltwater stream
is the Chicago River. Its natural direction of flow was out of the
Lake as recently as two thousand years ago, before a combination
of isostatic rebound and erosion at the outlet of Lake Huron cut it
off. (Lake Huron and Lake Michigan are connected hydrologically to

each other through the three-hundred-foot-deep Straits of Mackinac, and when the level of one lowers the other drops with it.) When the engineers of the Chicago Sanitary District attacked the river at the close of the nineteenth century, its source—in a great swamp at the present location of the city of Des Plaines, Illinois—was only eight feet above its mouth. It was no real trick to deepen the river eight feet and cut a canal through the marsh so it would drain into the Des Plaines instead. The engineers did not really build something new here, they just put something back.

I arrived in Chicago by commuter train, the 7:59 express from Naperville to Union Station. I was alone: Melody was spending the day going over some family matters with her cousin, Helen Beavin, and Helen's husband, Bart. It was an interesting parallel to the first trip, when I had also been alone in Chicago. Larry had gone ahead to Kalamazoo, Michigan: I had stayed behind to talk to a tall, slender, sandy-bearded EPA official named Kent Fuller. I was going to talk to Kent again today. With him would be Bob Beltran, an old friend whom I hadn't seen since Kent had hired him away from Oregon, with my grudging help, back in 1985. Toxicologist Milt Clark, whom I had not met previously, would join us partway through the morning to talk about fish advisories.

Union Station is on the west bank of the Chicago River, so the reversed river was the first thing I saw when I emerged blinking from the dark bowels of the station into the bright wind of the Windy City. I would like to be able to tell you that it was a thrill to see a river flowing backward, but in truth the flow was barely perceptible and in that flat terrain its direction seemed pretty irrelevant. I did notice that it had been cleaned up some. The cleanup had actually been started fifteen years ago, but all that could be seen were cosmetic changes and there weren't many of those: flowerpots here and there along the riverbanks were pretty much it. The Metropolitan Water Reclamation District (MWRD) was already hard at work, however, deep beneath the bowels of the city; and in May 1985 the first stage of TARP, the MWRD's Tunnel and Reservoir Plan, was put into operation. It was officially termed the Mainstream Tunnel, although everyone today—including the MWRD—calls it the Deep Tunnel. Three hundred feet beneath the city, it is thirty-three feet in diameter and thirty-one miles long, and it can hold roughly a billion gallons of wastewater. Sewer overflows which used to go into the river are now diverted to the Deep Tunnel and stored there until they can be properly treated. The change in the river has been dramatic. Gazed down on from the new riverfront plaza at Union Station, the Chicago actually looks like a river these days, instead of the open sewer it more or less resembled in 1983.

The Ralph W. Metcalfe Federal Building is five blocks east of Union Station along Jackson Boulevard. Nearby, in Federal Center Plaza, a farmer's market was going on under the tall, slim arches of Alexander Calder's bright red steel *Flamingo:* mesh bags of garlic hung from the posts of wood-and-canvas stalls filled with onions and tomatoes and scallions and sweet corn. I took the elevator to the Metcalfe Building's seventeenth floor and had barely begun to identify myself to the receptionist when Bob materialized at my elbow. He looked just as I recalled him from his Oregon days—blond and baby-faced, with thick glasses and a perpetually earnest expression that belie his status as an expert sea kayaker. Kent joined us a few moments later, still tall and slender, though the sandy beard I remembered had turned gray. We found a quiet corner.

The Chicago River, looking south from the Jackson Boulevard bridge in downtown Chicago, Illinois

"Be careful what you say," Bob warned Kent as I pulled out my tape recorder. "You never know who might listen to that tape." The tape I had made of my conversation with Kent in 1983 had been part of Bob's preparation as he headed to Chicago to interview for his original EPA job.

Kent reminded me that his role with the agency had changed. "Fifteen years ago I was a supervisor," he said, "in charge of environmental planning for the Great Lakes Environmental Program. I'm now 'Senior Advisor,' which means—officially—uh—'utility short-stop and grouchy old man.' But in something of the same perspective role, I think." He grinned. "So what from our immense store of wisdom are you interested in this morning?"

"Are the Lakes better or worse than they were fifteen years ago?" I retorted. "I think we can all run down the laundry list of changes, but what I really want to know is how you *rate* some of these things."

"Well, there's 'good news and bad news,'" said Kent, drawing a pair of sign-language quotation marks in the air to indicate the cliché. "The good news is that we've cleaned up the phosphorous problems, and we've cleaned up some of the egregious toxics like DDT. The bad news is that it's like the layers of a cabbage. We're exposing more and more layers. We've gotten into the whole realm of the 'gender benders,' the hormone mimicry things, and that's the big new chemical story. But chemicals are only part of the story, and in fact while we've been really concentrating hard on chemicals—which after all is a major mission of the EPA, as a good control agency—it's dawned on some of us that we've been ignoring another leg of the stool, and that's the whole habitat arena. The loss of habitat is staggering, and continuing. And you know, once habitat is gone, it's *gone*."

"Now, along the way getting to this point," Bob put in, "we've gone through a certain amount of evolution. We went through a phase where we and the Canadians worked together through the Water Quality Agreement to develop Remedial Action Plans. Then we followed that up with a phase of Lakewide Management Plans— the LaMPs. Another thing we have done is something called 'Mass Balance Studies.' The point of the Mass Balance Studies is to track the sources, pathways, and fates of organic and metal contaminants— which is, again, the chemical track. But what we've learned from that, in large part, is that we have a pretty good handle on controlling inputs of new contaminants to the system. There isn't much cranking down we can do. We've done graphs of toxic substances in the

Lakes over time. Back in the early seventies, in the bad old days, they were pretty high. And they dropped off. Now they're just kind of—wavering out there at a constant level. With the exception of toxaphene, which appears to be going up, and does seem to have some geographical concentrations."

"Where?" I asked.

"In relatively pristine areas," said Bob, "that tend to be—whether it's coincidental or not—downwind of pulp and paper mills."

"Ah," I nodded. "Whether it's coincidental or not."

"Understand that toxaphene was basically made by boiling turpentine and chlorine together," Kent said. "Turpentine is pinesap. And the fact is that the paper people are bleaching, uh—"

"Pine trees," said Bob.

"Pine trees. To make paper. Leads one to believe maybe there's a connection there. But I was going to back up just a little bit, a little while ago. In fifteen year's I've seen—well, for want of a better term, it's a paradigm shift. We're looking at biodiversity and habitat rather than just chemicals. I think one of the footholds for that was the revision of the Water Quality Agreement in 1987. The Agreement since 1978 had called for an 'ecosystem approach' to the Great Lakes, but there were"—he chuckled—"widely divergent views about what that meant. Well, Annex Two of the 1987 Agreement called for the naming of Areas of Concern where uses were impaired, called for Remedial Action Plans, called for identifying critical pollutants in the Lakes. Called for Lakewide Management Plans to address the impairments. And out of that there've been two schools of thought. There's one small hard-core group that says that it's a perversion of the Water Quality Agreement, it has no business in there, the Water Quality Agreement's essentially a *chemistry* agreement. It's water quality, and water quality means chemistry. Well, my view of the world is that there's sort of a three-legged stool here. One of them *is* the contaminants, the chemistry thing. One of them is exotic species. The third is habitat—habitat, and biodiversity. And you've got to look at all three to understand what is happening in the Lakes, and you've got to try to *deal* with all three to preserve and protect and restore the Lakes.

"There's tremendous developmental pressure along the shorelines for recreation. The major cities are poised for takeoff. The regional planning agencies here in northeast Illinois, and the Federal Reserve Bank, have looked at what they expect to see in terms of growth in the greater Chicago area in a decade, and they've got some years of it beginning to happen. We're going to add roughly the population of Philadelphia to this region in the next decade or two. And there's the aggravating pattern of urban sprawl. In the twenty years from 1975 to 1995 the population of the Chicago metropolitan area increased by about 4 percent. The shift from agricultural to urban land uses was *40* percent. And the sprawl has changed its nature, too. Used to be, you could pretty well get in your car and drive out to the edge of development. There's some of that left, but it's really splattered. I have some vacation property 360 miles from here in northern Wisconsin. There's a little subdivision that came in next to us. I went over to the first buyers, and talked to them—they were escaping their urban scene, because they couldn't live there. Where were they escaping from? A town about twenty-five miles away, with about two thousand people. So it isn't just people fleeing the big urban core, it's people splattering all over the landscape because they want to live with Mother Nature. But an awful lot of them, to come back to the point, are concentrating on the shores of the Great Lakes."

Bob nodded. "There's a common phenomenon of people building cottages in nice places along the shore," he pointed out. "And typically what happens is, when people reach a certain age, they say, 'OK, I think I'll retire to the cottage.' Now the cottage becomes a home. So they rebuild the place. They make it a year-round residence. And from that point forward it's no longer a cottage. It becomes part of the basis of a new town someplace."

"And it's not just that the septic tank is stressed," said Kent, picking up the thread again. "It's that they now have to have a sewage collection system. And they say, 'How are we going to fund that? Well—let's just OK a little more development so we can spread those costs. And let's go to a smaller lot size.' Fifty years ago the vacation place was a vacation place, and the kids would pick it up from their parents, and it would *stay* a vacation place. But now, as Bob says, the vacation places of the thirties, forties, fifties, sixties, and even seventies are becoming year-round residences. You're getting the streets—you're getting the infrastructure. The whole recreation area is becoming urbanized."

"Do you think that's related to the aging of the baby boomers?" I asked.

"Momentarily," said Kent. "But I think it's really communication and transportation. It's getting easier and easier to go live over by the Lake. You've got cable TV, you've got your internet connection, you've got good telephone service—and you can get on a high-way and come downtown for a play on the weekend if you feel like it."

"It's also related to affluence," Bob put in. "Years ago that didn't happen so much because, in order for people to live efficiently within their means, they pretty much had to stay where the services were. Now people can afford to have the services brought in."

"One of the things I noticed up in Door County," I remarked, "was the death of local businesses. The whole area is now either national chains or boutiques."

"Yeah," Kent agreed, and Bob said, "It happens everywhere, you know."

I nodded. "What you're saying is, it's not something that's iso-lated anymore to places like Door County, which have always been popular."

"No," said Kent. "It's going on all over the place. I just had the pleasure of spending three weeks in England. And I was really struck by the fact that, while the pressure is obviously there, they're dealing with it. They've got a program of supporting traditional agriculture—keeping the hedgerows, keeping the sheep in the economy. We said good-bye to animals and hedgerows a long time ago. We said, 'Get those damn hedgerows out of here. We've got a big tractor—we're going to corn and beans. Get the cows out of here. We don't want pasture, we want cash crops. And it's my land, and when I'm ready to subdivide it, and grow my final crop of houses, that's my right absolutely, and I'm gonna do it.' And we aren't going to change that, much. But there are some lessons to be learned where people in other countries are dealing with it." He smiled. "We've got a heck of a laboratory experiment coming up right now in northwest Indiana. Which used to be the pollution sink of North America. I mean, just a dreadful place—"

"You're talking Gary to Michigan City."

He nodded. "Gary—Hammond—South Chicago—all around. And the cleanup's there. The rivers used to be nothing but sludge worms, and now they've got goldfish in 'em, and bluegills, and sunfish coming back. And a *huge* population pressure is going to hit that, as that combined growth and sprawl we were talking about continue to hit the greater Chicago area. How are they going to handle that? Their environment's coming back. Biologically, it's very, very rich. There are more species in the Indiana Dunes National Lakeshore than there are in all of England. Will they build on that? Will they be able to perceive the importance of the natural resource, and use that as a foundation for sustainable development? The jury's out. We'll see how it goes."

"One of the positive things that has been striking me," I remarked, "is that, where there was a sense last time of cities turning their backs on the waterfront and developing inland, and just using the water for trash, that doesn't seem to be the case anymore."

"Yeah," said Bob, "that's been a major turnaround."

"The first Mayor Daley," observed Kent, "used to dream of kids catching fish out of the Chicago River, and everybody, including me, laughed. The next Mayor Daley is saying some of the same things, and people are paying attention. It's a possibility."

Bob nodded. "I think the Remedial Action Plans have been instrumental in much of that," he said. "As soon as people start thinking in terms of—'Hey! The water's really getting cleaned up!'—it pushes up land values and changes what people want to do with the waterfront. Once you have economic interests focused on the waterfront, it changes the whole impetus. But Remedial Action Plans are a business that this office got out of."

"Everybody seems to be getting out of that," I pointed out. "The states have cut funding for them—"

"You know, that's interesting," Kent interrupted, "because I really think that the Remedial Action Plans are one of the very best locally based initiatives that we've managed. And I think that's the long-term promise. Having a national law that says you've got to do something at a standard rate across the face of the land was valuable to get the egregious stuff under control. But it doesn't answer the local problems, where you may need a good deal more. This agency has never understood locally based initiatives, and never will in my opinion, although we've got a mantra that goes, 'locally based'— 'community-based programs'—uh—"

" 'Community-Based Environmental Planning,' " said Bob.

"Yeah."

"We have a team for that."

"Yeah, yeah, I'm sorry." Kent protested. "I'm getting close to retirement, so, as I say, I'm working on being a professional grouchy old guy. But I've also spent thirty years in local government, on my own local planning commission, as an elected official. So I think I know something about local things. And that's why I've been so damn interested in RAPs all along. I think federal agencies can nurture these kinds of programs, but they've got to be locally based. The RAPs were a huge step in the right direction, and I really mourn the fact that this agency hasn't been of more help to them."

"One of the people that I talked to fifteen years ago," I said, "was Jane Elder of the Sierra Club. Jane, to most people's eyes, has sort of dropped out of sight. I talked to her in Wisconsin a couple of days ago. She's now working on a project to figure out why the public can't grasp biodiversity as an issue."

"Time," Kent said immediately. "When I joined EPA, I was warned —somebody said, in a moment in an argument, 'You sound like an environmentalist.'" He laughed. "I mean, it was a bad thing. You know, in the EPA, you were supposed to be engineers, or real estate people, or something. And I as an environmentalist was suspect. The term 'ecosystem' was not understood. I was told to take it out of things that I wrote—not adversarily speaking, but because nobody knew what it meant. And in twenty years, the term has become commonplace."

"But does anybody yet know what it means?" I asked.

"Yeah, well, we could spend some time on that," agreed Kent, "but at least the term is familiar to people, and the concepts. Of course, most of the people who watch the Discovery Channel are going to have a much clearer idea than those who don't. But it's entered the common parlance. I think in twenty years 'biodiversity' is going to be also much more into the stream. The kids are being taught about it in schools. I'd love to see what Jane comes up with, and I'm sure there's real puzzlement—but I also think we just need time for it to come into general use and understanding. And I think the use of the term 'ecosystem' is a fair comparison."

"There are other factors, though," I said. "One thing Jane said was that she's found that people don't identify with the Great Lakes as a region. Her sense is that it's too large an area, and too diverse an area, to identify with."

"Yeah," Kent agreed. "We've bumped into that in terms of the Lakewide Management Plans, in trying to get people interested in 'The Lakes.'" He gestured with both hands. "They're so vast, it's hard to get people to think about doing something to help them. And I think that was the strength of the RAPs. Because now we had a unit that people could focus on. 'The harbor.' They could be concerned about 'The harbor'—they could think about doing something for 'The harbor.' You need to build around something small enough to grasp. I think Jane's absolutely right. People *don't* see themselves as part of the Great Lakes system."

"People relate to the Lakes as that part they can drive to within half a day," said Bob. "They have difficulty relating to the people on the opposite side. The Lakes are truly vast. We have a problem in trying to deal with other people in this agency, especially on a national level, because they have no concept of how vast the Lakes are. But people who live on the Lakes have no concept of how vast they are, either."

"Is it any different in the Great Lakes than it is, though, for environmental awareness across the country?" mused Kent. "I bet I could go down my street and find fifty people, and ask them all what river basin they live in. And they wouldn't be clear. They wouldn't be sure whether it was the Chicago River, or the Des Plaines River, or something else. I think they occasionally take visitors over to see 'The Lake,' and say, '*Gwarsh, yuh can't see across it!*' But how many people have canoed the river that runs through their own suburb? Not very many."

"You mentioned the Des Plaines River," I said. "We're talking about the difficulties of people seeing themselves as Great Lakes residents. If they live on the Des Plaines, they actually *aren't*, technically. But how far out do you go?"

"Well," said Bob, "there are many layers. If you're talking surface water hydrology, the Great Lakes Basin is relatively easy to delineate. The Corps of Engineers has done it, and they have maps that show what it is. If you want to talk groundwater hydrology, you'd have to develop a different map. If you want to talk ecosystems, that's another map. If you want to talk human society, that's an entirely different map. But when it comes to diversion of water, there is a legal definition of what's considered the Great Lakes Basin, and I don't know that we're going to make any changes in that."

Kent was looking amused. "Wasn't it Charles Dickens who said the law is an ass?" he said. "I mean—you start getting into legalisms, and it's always just goofy. How do you deal with Chicago? Do you take currently? Well, the sewage collection systems take virtually all our sewage down the Mississippi River. Do you take historically? The Chicago River went into Lake Michigan. Or, do you say, 'Wait a minute, we've got an air plume from the Chicago complex that spends a lot of time out over Lake Michigan, dumping stuff there. Should we look at that?' It's a huge, interrelated piece. Pragmatically, I think everybody clear out through Elgin is part of the Great Lakes ecosystem, because they come downtown. And if they pee here, it goes—well, it would have gone into the Lake, but now it goes down the sewage treatment plant. I don't know." He fell silent.

"Yeah," I agreed. "There's a real question in my mind, whether you're talking about what you're taking out of the Lake, or what you're putting in. If you're talking about taking out, as you said, when anybody comes downtown and pees it goes into the Mississippi. If you're talking about what you're putting in—"

"The car exhaust goes in," said Kent.

"Are you familiar with the recent controversy over the bottled water being sent to Japan?" Bob asked. "I thought that was especially interesting. Proportionate to the water that gets taken out and pumped into municipal water supply systems outside the basin— you know, that's just such a tiny token of a drop."

"That's already being done," I pointed out. "It's just that they put hops in it first."

"Yup." Kent laughed. "That was the Wisconsin position—as long as it went out in beer cans, it was all right."

"Yeah, that occurred to me, too," Bob agreed. "Not to mention Coca-Cola, Pepsi-Cola, and—"

"Well, they move syrup around the globe, and then they use local water," said Kent. "But the beer goes out all over." He laughed again. "I don't know—I have a lot of trouble taking that whole debate seriously. It really is an issue of whether New York gets to put that bucket of water through the turbines at Niagara Falls to generate electricity, or whether Minnesota gets to sell it. It's a lot of economic arguments, really. The thing to think about, if you want to worry about where the Lakes are going in terms of levels, is climate change. And that's *going* to happen, and *almost* certain to have a huge impact."

"Some of it is already happening," said Bob.

Kent nodded. "The die is cast. The Lake level hasn't changed particularly in response to it, but the models *do* say it's going to get warmer. It's probably going to rain more, but it's going to be so much warmer that we're going to get more evaporation than we will get rainfall, so the Lake levels will drop. But the models are still pretty rough, and they're still guessing."

"Actually," said Bob, "it may be that the effects are already being felt. Back in 1958, the U.S. and Canada developed Plan 1958-D to manage the level of Lake Ontario. They set targets for Lake level control under various scenarios, but they also provided for deviation from the formula if conditions warranted. Plan 1958-D has been deviated from every year since 1958. And the deviation has almost always been on the side of trying to hold water levels down, because they've been higher than anything that had been experienced or projected previously."

"It's likely to go the other way with climate change, though," said Kent. "And the real impact's on the St. Lawrence River, because they can control the amount that they let out of the Lake. They can hold the Lake level."

"But one of the things they did when they created the St. Lawrence Seaway," Bob persisted, "was that they created Lake St. Lawrence. And Lake St. Lawrence has become an entity in its own right. So now one of the forces that resists allowing more water out of Lake Ontario is that that raises water levels in Lake St. Lawrence. So you have a whole community *there* screaming and yelling. And then there's the whole Power Authority, and *they* don't want to see anything going through that they can't put through the turbines."

Kent nodded agreement. "And then you can add to that," he said, "that a major adverse impact on wetlands in Lake Ontario is the lack of Lake level changes. Because they hold the thing in such a narrow zone that they're not drowning out the invading trees and brush in wetlands. You know, wetlands are dynamic. They're a stress zone, and it's because of that stress that they exist. And they're withdrawing the stress, and—things suffer."

"Yeah," said Bob. "Lake Ontario, under the plan, is to be held within a certain range that they will not go above or below. Well, the net effect of the way they've been regulating it is to hold it in the *upper half* of that range almost all the time. So the Lake's hydrologic system is not experiencing the full effect."

"Well, the whole physical Lake level thing is largely a habitat-related issue," Kent stated. "There's a local project I'm having fun with. It's called 'Chicago Wilderness.' Lovely oxymoron. I'd like to spend a little time on that—if I might." He looked at his watch. "I have to get back to my desk pretty soon."

I nodded. "Go ahead."

"Good." He shuffled through his briefcase. "A couple of years ago some of the leadership in some of the major organizations said, 'Look, a lot of us are touching biodiversity, but it really is not anybody's principal mission.' So a handful of us got together, ad hoc, to see what we could do. And the result was that we formed the Chicago Region Biodiversity Council. Then we went through a lot of debate, and we said 'Chicago Wilderness' is going to be the name of this effort. Come *on*, 'Chicago Wilderness,' what kind of crazy—but 'Chicago Region

Biodiversity Council' doesn't exactly roll off the tongue and fix itself in the memory."

I laughed. "No."

"It's got some baggage with the Illinois collar counties and Indiana, but it's *such* a good hook, in terms of what people react to. We wanted to tell the story of what the biodiversity of this region is, to help people realize where they live, biologically. So we did. We created an *Atlas of Biodiversity* for the Chicago region." He placed a glossy bound document before me.

"We're trying to look forward," he continued, as I leafed through the *Atlas*. "How do we solve these problems, and help people grasp—"

"Basically, how do we answer Jane's question," I supplied.

"Right. How do we get people to know where they live, and what they can do about it, so they don't just go to the bird sanctuary and see if they can see a cardinal. I've lived almost my entire life in this area—I went to school in Pennsylvania, but I was born and raised here. And I live within two miles of a 'County Forest Preserve.' To me, that was a place where there were drunks and mosquitoes and poison ivy. Nature was in northern Wisconsin. I got involved in a volunteer stewardship program with The Nature Conservancy— *damn!* This miserable poison-ivy place within two miles of my home is *rich* in biodiversity, it has an *incredible* spring wildflower flora—it's an old savannah remnant, with some maple woods in it—it's terrific. I didn't have a clue. I've got a degree in forestry, I work for the U.S. EPA, and *I* didn't know anything about where I lived. Well—hey, we ought to be able to do better than *that*."

"Maybe we can get the drunks out," I said.

"Ah, hell, they can stay, as long as they don't trample the rare species." We all laughed. "Don't hassle the people who want to use the place. But I do think we need to find ways to help people understand where they live, and the fact that they can make a difference. The whole thesis of Chicago Wilderness is that this is a *remarkable* place. And the people ought to enjoy it. They ought to have as much a feeling that this is a special place as somebody who lives near the Grand Canyon."

I asked how nonindigenous species fit into the picture. Kent looked thoughtful. "It's a huge problem," he said, "and I don't know how we deal with it long-term, because with global transportation, I don't know how we keep everything out. The premier bad guy in the Great Lakes is the zebra mussel. It, and a variety of other things, came in ballast water. And it's really *stupid* of us to have let that happen. Frankly, when you look at the cost/benefit of the St. Lawrence Seaway, it's been a bust. We probably haven't made much in hard dollars, and we've lost an enormous amount ecologically. And once the exotics are here, there's not much you can do. We've got purple loosestrife decimating our wetlands—it's all over the place. We've got things like garlic mustard and sweet clover that we battle all the time. Eurasian buckthorn. Buckthorn's a real misery, and it left all its predators wherever it originated from. You look at the leaves in the fall, and they don't even have *holes* in them. Nothing eats the crap." He sighed heavily. "I digress. The point is, stuff's going to come at us, and I think we've just got to be sensitive to it, and not let it take over. Of course, 97 percent of the stuff that comes in is *not* a problem. I do prairie walks on weekends now. I've got a place where I take people in. There's seven-foot teasel, which is a real bitch. That stuff is *mean*. And I say, 'OK, here's the teasel—here's the loosestrife—but now, here's chicory. Here's Queen Anne's lace—here's a couple of other things. They're welcome visitors.' They mix right in. It's that small minority of invasives that somehow we've just got to do a better job of trying to get out and keep out."

"So what do you do about something like the zebra mussel?" I asked.

"I think you look for biological controls," said Kent. "I'm not really equipped to speak, but my impression is that when you get a species coming in, it flushes like crazy. We got earwigs here ten years ago— I'd have to shake my washcloth out in the morning before I could wash my face, because there'd be earwigs in it. And you can hardly find one now. So they flush through."

"Yeah," I said. "Some sort of control develops."

"Well, the centipedes exploded in my neighborhood. And I'm sure a lot of other things happened. The point being—the Great Lakes has the world's richest collection of freshwater mussels. It may be that where we've got an unusual species, or a species of special concern, we'd better manage some refugia and try to keep viable populations protected until the surge goes by, so that the wave of zebra mussels doesn't wipe these things off the face of the earth. That's hardly a big answer, but maybe it's something we should be doing."

"Bob was making a point earlier that there's still a dissolved oxygen problem in Lake Erie," I said.

"Arguably there was presettlement," Kent said. "One of some sort. But go ahead."

"I was relating it to zebra mussels," I said. "I was told yesterday that zebra mussels develop an oxygen sag around themselves, because they use oxygen so voraciously."

"Well, they certainly change things," Kent said. "I mean—right now, my drinking water tastes like bread mold. I mean, it tastes just awful. It's a chemical associated with the algae that are favored by zebra mussels. So—all *kinds* of impacts. I've picked up on the amphipod populations lately in Lake Michigan. There's these huge dead zones, now, where they can't find *Diporeia*. It was the predominant food supply for some fish, at least in some life stages. And it's just gone."

"Yeah," Bob agreed. "We could probably wax eloquent without a lot of knowledge, but there's a lot going on in southern Lake Michigan in terms of currents, and something called the Lake Michigan plume—and the EEGLE study which is going on between us and NOAA—but before we—Oh, there's Dr. Clark."

Milt Clark burst energetically through the door. "Oh," he laughed, "we get the big introduction."

Bob grinned. "Yes. Milt Clark, this is—"

"I'm Milt Clark," he said, extending his hand. He was dark-haired and nattily dressed, which is to say he was wearing a sport coat. The rest of us were in shirtsleeves.

"Well, Clark's going to talk about chemicals, and I *know* they're of no importance to the system, so I'm going to excuse myself," said Kent loudly.

"He's had to hear me for too many years," Milt smiled.

Kent laughed. "Habitat's the only relevant issue."

"Thanks, Kent," I said. "See you in fifteen years."

"Sure." He disappeared.

Milt had come to discuss fish advisories, a topic that has been pitting the EPA against the state of Michigan. The agency has been trying to establish uniform Great Lakes advisory standards; Michigan has been vehemently resisting. "Each state used to have a different fish advisory program," Milt explained, "and if you got to Lake Michigan that meant you had four fish advisories out there on one Lake. So we came up with a proposal to establish a uniform protocol for fish advisories based upon a lifetime cancer risk of one in ten thousand. At limited fish consumption, this would be at levels of .05 parts per million or less.

"Michigan was resistant to that, but the resistance really came, not from the health agencies, but from the Department of Natural Resources. They were concerned about, 'Well, if we have too many restrictions on the fish it's going to cut into recreation and tourism, and aren't we being overly protective?' We kept sending out letters saying, 'States, please move this forward,' and Michigan was going the other way. And when it came to a decision on salmon, they ended up rescinding the current advisories. We kept writing documents and communications, challenging their conclusions, and then eventually, in 1996, the Jacobsons published their work on Lake Ontario showing six-point IQ deficits in children at age eleven—"

"I'm an example of that, by the way," remarked Bob, who grew up in the tiny town of Sandy Pond, at the eastern end of Lake Ontario.

Milt grinned. "Anyway, all this stuff came out, and we wrote a report saying we thought people were at risk. We sent it out to all the Great Lakes states and said, 'Please move forward.' Michigan still held out. We sent a couple of letters saying, 'If you don't do this to protect women and children and other segments of the population, we will.' And in collaboration with the Agencies for Toxic Substance and Disease Research in Atlanta we had 1.2 million fish advisories printed, and we sent them out to the places where they sold fishing licenses. We would have liked to send them directly to the anglers, but it was not—let's say—*easy* for us to get a list of anglers from the state." He smiled. "It took us quite a bit of effort to get the places where they distribute the *licenses.* And we kept hammering our message that we still needed to have these advisories in place. The governor's office reconvened their scientific body—the scientific body said, 'By golly, this .05 value looks good to us for women and children—' "

"Fancy that," I said.

" '—so we think what we're going to do, is, we'll protect women and children, but we'll let adult males and everyone else do their own thing.' And after we looked at the situation—there'd been extensive media attention—we decided, even though we want everyone protected at the same level, we'll accept this as an interim decision of the state. The state of New York has since sort of followed suit with Michigan. So we now have six of the Great Lakes states protecting everyone at the same level, with Michigan and New York protecting women and children like all the other states but having a deviation for other segments of the population. So that's my knowledge." He laughed. "I'm drained. That's all I know."

"He's an empty vessel now," smiled Bob.

I asked how the EPA intended to deal with the fact that up to 60 percent of women who eat Great Lakes fish are still unaware that there are advisories on them. "We're going to try to gear up to get this out to more doctors," Milt said. "More county health departments—more schools—there's going to be an effort to do expanded distribution." He chuckled. "Of course, what we're really going to do is clean up all the Great Lakes and eliminate the pollution."

"All we have to do is remove all the contaminated sediment," said Bob. "The hundreds of thousands of square miles of contaminated sediment—"

"Well, nail the rivers," said Milt. "I mean, we could do a lot with a couple of billion dollars. That's about what we need. But really, the thing we always get to, that we've all talked about, is a holistic approach to the Great Lakes. We've got this chemical component, and Kent's issue of habitat destruction—all these things, all going after the ability of the Lakes to heal themselves. I frankly look at it as sort of a multiple-stressor hypothesis. And as scientists we tend to get the little laser beam. But how do you implement a holistic approach? We get work groups, we get everybody excited here, we start something like this and we build up for a year or two, and then it—goes back down."

"Well, the fact is," said Bob, "that, in a practical sense, in order to deal in a holistic manner you have to have a lot of people working on small pieces. No one person can deal with everything. So you have a lot of people dealing with little elements, and it tends to put the blinders back on. You have to keep working at getting people to wake up and say, 'Oh, wait a minute, there's the rest of the world out there.' "

"That's right," said Milt. "I don't know how Bob feels about this, but the EPA—a couple of years ago, you know, we almost got shut down. We've had to undergo major reorganization. And frankly, I think we have to run a mode of reenergizing again, out of what has been a little bit of pain. I believe you've got to have certain defined goals out there, and you've got to work toward those goals. And I don't know how much we're doing that any more. They keep saying, 'Here's the goal for the year 2000,' and we're doing that some—we're doing it with PCBs and mercury, we've got international agreements to seek 90 percent reduction of these things. But at the same time, you've got this situation now where there's amendments, and riders, in Congress. You know—'No, we're not going to do any dredging of these rivers until the National Academy of Science can take a look at this, and we've got to stop this mercury plan that EPA's got, and restrict this.' When we really *do* try to set some goals out there that will make a difference, everyone starts screaming, because it's gonna cost some money."

"Yeah," said Bob morosely, "unfortunately, sometimes science can be used as a barrier as well as an energizer."

"You know," said Milt. " 'We need more science.' 'Study it more.' And—"

"Of course, that's something a lot of people have held against us as an agency," Bob interrupted. "We're always saying we need to do more studies. And from a science standpoint, often you *do* need to do more studies. You can't just charge in. But that can also be used as a way of—throttling back—agencies or other people who think that they *have* enough information."

"There's never enough data," I said.

"Well, that's becoming almost a new sort of buzzword ploy," Milt responded.

"Yeah," said Bob, "the buzzword thing."

" 'Study it further.' " Milt drew exaggerated quotes in the air. " 'More science.' "

"Yeah," Bob and I chorused.

"Or 'Good Science.' "

"Right. Right."

"But that just becomes another tool to—you know, here we have *stacks* of reports, external peer reviews of all of this, and there still will be some little area in there that they can find. And if they can't attack it, then they go to Congress—"

"I'm sure there's somebody out there who has work groups dedicated solely to finding something that they can delay, and base it on," said Bob. "But when you hear the buzz terms now—it seems like we have an ever-increasing rate of buzz-term creation."

" 'Good Science,' " said Milt.

" 'Quality Science,' " said Bob. " 'Science-based.' Most times those are really words for—"

" 'Let's not do it,' " I supplied.

" 'Let's delay,' " Bob nodded. " 'Let's not do something.' " A pall seemed to be settling into the room—one that I was to see duplicated many times over the next several weeks. Scientists, especially those in the agencies, are worried that science is being marginalized these days, and undercut, and not listened to. The "Republicanization of the Lakes" that Jane Elder had complained of is being manifested in three ways. Good science is underfunded. The results of good science are ignored or ridiculed. And junk science—science undertaken not for knowledge, but to further an agenda—is taking over public discourse and being pushed by legislators who, in an ironic sort of saving grace, won't fund the junk science either.

We spoke briefly of the enigmatic yellow perch decline in Lake Michigan. "Nobody knows as much about it as we would like," Bob stated. "Hypotheses are what we've got. We've got some evidence, but I don't know if we have anything relevant. The yellow perch decline may be related to things like the round goby. May be related to the zebra mussel. May be a compound of zebra mussel and goby. May be related to the fact that gobies eat zebra mussels, and large perch eat small gobies. Gobies, as I understand it, can pass zebra mussel shells whole through their bodies. Perch, I'm told, can't. One hypothesis is that the perch eats a round goby that's eaten a zebra mussel, and it kills the perch. There's talk of habitat competition between the perch and the goby, but there's some other evidence that says they don't use the same habitat at the same time of year in the same way. So I don't know that I can give you a lot of information, but I can give you a lot of speculation."

"There's a sex change of some type," said Milt. "And there's some size differential in the fish, and that's all I know." He and Bob speculated that this could be related to the endocrine disrupters—what Kent had called the "gender benders"—that have lately been dominating pollution-control advocates' concern. It was nearly noon. I had been talking with the EPA for two and a half hours.

I HAD A LUNCH DATE WITH JOHN WASIK, special projects editor for *Consumers Digest* and the author of *Green Marketing and Management: A Global Perspective* (Blackwell, 1996), whom I had met at a conference of Great Lakes journalists in Detroit a year before. I invited Bob to come along. It seemed a shame to spend a whole morning with someone I hadn't seen for ten years and not talk about anything but work, and I didn't think John would mind. Fortunately, he didn't seem to. We met him at Cavanaugh's, a combination restaurant and microbrew bar on the ground floor of the Monadnock Building. The Monadnock is sixteen stories high, and although it looks puny in its current company—the Metcalfe Building is next door, and the 110-story Sears Tower is only a few blocks away—when it was completed in 1893 it was the tallest office building in the world. The restaurant was a world of elegant old wood and earnest young waiters, and it was popular enough with the downtown Chicago lunch crowd that they wouldn't let two people occupy a table for three: Bob and I weren't allowed to sit down until John arrived, perhaps five minutes after we did. The table they finally took us to was crammed against a thick inner partition near the center of the building, looking toward the east wall. The windows in the east wall were small and deep, like arrow ports in a castle. The Monadnock was built before William Le Baron Jenney had finished inventing the steel-framed skyscraper, and the bottoms of its bulwark-like outer walls are sixteen feet thick.

"I'll take the seat with its back to the wall," I said, after a moment of hesitation.

John, whose prematurely silver hair gives him a passing resemblance to comedian Steve Martin, nodded sagely. "Always a good idea in Chicago," he said.

Bob regaled us with stories of kayak trips with his wife through the Pukaskwa, more than one hundred miles of wild Superior coastline in Canada, where polished walls of stone rise more than a thousand feet directly out of the water. Waterfalls descend over them; storms send forty-foot waves crashing against their feet. "The highest point in Ontario is Tip Top Mountain in the Pukaskwa, only a few miles back from the Lake," Bob told us. "There's a point almost as high on Michipicoten Island, ten miles offshore. It's the most spectacular landscape in eastern North America."

The talk turned—as it often does among those who love the Lakes—to the great dearth of understanding about the Great Lakes, even on the part of people who may hold the Lakes' fate in their hands. Bob told us of an official at EPA headquarters who mentioned to Bob's former supervisor one day that he'd like to come to Chicago and see Lake Michigan—and maybe jog around it. John and I were incredulous. Bob said his former supervisor insisted it was true. "But it also goes the other way," he pointed out. "Activists don't really know the Lakes, either. They keep insisting that the Lakes are filthy, and in many places they are. But that's just along the shorelines. If you get out in the middle of them, you can drink to your heart's content—even Lake Erie. Despite all that we've done to them, the Lakes are in better shape than we could ever expect—or maybe have a right to."

After lunch I walked down East Jackson to Grant Park, at the edge of the big blue Lake. Boats in the Chicago Yacht Basin bobbed gently up and down on small swells that had made it past the breakwater. The yacht basin lies just north of Northerly Island, which houses Merrill C. Meigs Airfield: if you're a fan of Microsoft Flight Simulator, you have probably flown over the basin—in virtual space—many times. Actually, you are more or less in virtual space as you stand here. Grant Park is built on fill, the remnants of pre–Great Fire Chicago shoveled into the Lake after Mrs. O'Leary's cow leveled the place. The city fathers' early plan to stuff it full of buildings was thwarted by Montgomery Ward, the founder of the department store chain that bears his name, who waged a vigorous one-man battle for many years to ensure that the area lakeward from the old shoreline—at Michigan Drive—was kept as open space. Critics claimed he did it to keep his own view of the Lake from being blocked. Perhaps they were right, but I think we can live with that brand of selfishness. Chicago today is the envy of many cities that didn't have a Montgomery Ward to fight for their waterfronts.

After I had been in Grant Park for a while, I began to notice something distinctly odd. There is a cement walkway along the top of the seawall that runs the length of the park, and small knots of tourists were gathering there every few minutes, taking pictures of each other. Almost always, the person with the camera was standing by the Lake and looking toward the city. For most Chicago tourists, it appears, the Lake isn't even a backdrop—it's merely a convenient open space whose shore lets them get back far enough to see the skyscrapers. The blue water lapped and the boats in the yacht basin bobbed and the gulls screamed by, and the tourists shot pictures of each other standing against big buildings. The human body is 78 percent water. I wonder why we continually do as much as we can to deny that basic connection?

I wandered over to the Buckingham Fountain, called the park's "centerpiece" by most tour guides, though that again ignores the Lake. The fountain was built in 1927 of red marble imported from the state of Georgia. It is immense: 128 feet across, with a central jet that sends water 135 feet straight up into the windy Chicago sky. I thought about our lack of connection to the places in which we live. Importing marble from seven hundred miles away to build a fountain must come from the same strange place in the human psyche as does photographing yourself against buildings instead of the Lake. They are both a denial of our bond to geography. We don't interact with systems as systems, Jane Elder had observed, and it is true. So how do we—as Jane had also observed—get the American people to sit up and pay attention? Attitudes are better—truly—than they were in 1983. But as long as a man-made fountain is declared the water centerpiece of a park on the shore of Lake Michigan, we have come nowhere near far enough.

THE FIRST SERIOUS ATTEMPT TO CREATE A national park in the Indiana Dunes can be credited to Steven T. Mather, the first director of the National Park Service, who began agitating for a Sand Dunes National Park as early as 1916—the year the Park Service was created. As with Montgomery Ward's waterfront, there were whispers that the director's motives weren't totally altruistic: Mather was a Chicago businessman, and the proposed park was almost literally in his backyard. Still, there were reasons. This land of bright dunes hovering between blue sky and bluer Lake has been recognized as a special place for a long time. In 1913, three years before Mather's efforts began, a group of European scientists preparing for a trip to the United States sent ahead a list of the places they wanted to make sure to see. The list had four names on it: Yellowstone, Yosemite, the Grand Canyon—and the Indiana Dunes.

In 1966, fifty years after Steve Mather put it on his short list of parks to create, legislation establishing the Indiana Dunes National Lakeshore finally passed Congress. By then it was almost too late. The destruction of the Dunes had begun long before Mather got there, with trainload after trainload of sand sent the eighty miles around the corner of the Lake to Chicago in the aftermath of the Great Fire to replace the buildings whose remains now lie under Grant Park. Soon after that, industries began spreading toward the Dunes, reaching Burnham in 1870 (Illinois Steel), Whiting in 1889 (Standard Oil), and East Chicago in 1902 (Inland Steel). In 1906, U.S. Steel established a company town in the western end of the dunes belt and named it for the chairman of their board of directors, Chicago attorney Elbert H. Gary: leveling and filling nine thousand acres of dune-and-swale topography, the creators of Gary displaced more earth than was moved a few years later to build the Panama Canal. At roughly the same time, a group of businessmen in Porter County, Indiana—hoping to create a Gary of their own—short-circuited the

Little Calumet River directly into Lake Michigan through a straight canal with the unlovely name of the Burns Ditch. A canal with an even less lovely name—the Calumet Sag Channel—diverted much of the Grand Calumet River into the Chicago. In the early 1960s, as the battle to preserve what was left of the Dunes reached its climax, Bethlehem Steel—which had quietly been buying up land for a number of years—began tearing up the heart of the proposed park to build a massive foundry. The Northern Indiana Power Supply Company (NIPSCo) and Midwest Steel owned large amounts of land near Bethlehem's site at the mouth of the Burns Ditch, where Porter County boosters were still pursuing their fifty-year-old dream of a Port of Indiana. In 1966, the same year legislation creating the Indiana Dunes National Lakeshore finally passed, federal money came through to construct Porter County's port. Midwest and NIPSCo immediately broke ground for large facilities.

As the industrialization of the Dunes proceeded, so did a destructive force of another kind. The railroad tracks that took mined sand and milled steel to Chicago also ran the other way, bringing Chicago residents out to the Dunes to play. Some of them stayed. Housing communities—Dune Acres, Ogden Dunes, Beverly Shores—sprang up. The South Shore Line began offering regular commuter service into the Windy City.

The communities and industries and artificial waterways that lace the Dunes have resulted in a desperately fractured natural landscape, and the boundaries of the National Lakeshore reflect this. On a map, the park looks like spilled paint: a couple of large puddles, a splatter of droplets, a long dribble line stretching out from one of the puddles toward the other as if the can had continued to spill while being righted. The Lakeshore doesn't even protect the highest and wildest of the remaining dunes: the state of Indiana does that, in a state park—established in 1923—which the federal preserve wraps around in the same awkward manner as it does the industries and the residences and the power plants. Cooperation between the two parks exists, but it is far from perfect. Given their varying missions, the state park emphasizing recreation and the federal park emphasizing preservation, it is doubtful that truly excellent cooperation can ever be achieved.

Hemmed in and cut to pieces on the landward side, the Dunes are also shrinking away on the shoreward side. At the park's eastern end, in Michigan City, there is a harbor and small-craft marina protected by a groin called the East Pier. Groins are walls built vertically out from shore to trap the sand—known as *shore drift*—that moves laterally along a shoreline. This prevents harbor entrances from developing bars across their mouths, dramatically decreasing maintenance costs. Unfortunately, it also starves down-current beaches. In Lake Michigan the prevailing direction of shore drift is clockwise, from Michigan City into the Dunes; so ever since 1904, when the East Pier was constructed, sand that should have nourished the beaches of the National Lakeshore has been piling up behind the pier instead. The situation was compounded further by the construction of a giant NIPSCo generating facility beside Michigan City's harbor in 1972. Together, the protective works constructed for the plant and the harbor eat away at the federal and state parks and threaten the homes in the dunesuburbia communities around them. The closest houses to the water have become much closer than their owners want. Some are close enough they are actually beginning to fall in.

NIPSCo's Michigan City power plant from the summit of Mt. Baldy, Indiana Dunes National Lakeshore, Indiana

In 1983, I was shown around the Lakeshore by a Park Service naturalist named Robert Fudge. Fudge was long gone in 1998, having climbed the federal ladder to the position of assistant chief of the Park Service's Division of Interpretation and Visitor Services in Washington, D.C. Ron Hiebert, a research administrator for the park whom I had briefly interviewed, had moved on as well. Melody and I had two choices: see the National Lakeshore in the company of a recent Park Service transplant to the area, or see it unguided.

We decided to see it unguided.

The spell of the Dunes was apparent as soon as we left Interstate 94, the freeway that hems them in on the south. The two narrow lanes of U.S. 12 bored through a dim green tunnel of hardwoods, past marshes and wooded homesteads and tiny railway stations. From the Dorothy Buell Visitor Center, Kemil Road took us north to the beach; soon we were frolicking in big Lake Michigan breakers under a perfect azure sky. Far to the west the tiny towers of Chicago, barely visible at the skyline, seemed placed in proper perspective to the blue immensity of the warm, wide, wonderful world of sea and sand.

Comparing pictures today, I can see that the beach at Beverly Shores was perhaps a third again as wide in 1998 as it had been in 1983. That is not what I noticed while I was there, though: what caught my attention were the people. The park has been popular from the beginning, with visitor numbers exceeded only by those of the big flagship parks such as Grand Canyon and Yellowstone. At the time of my earlier visit, however, most of that use seemed to be concentrated at the ends of the Lakeshore, at West Beach—which functions more or less as a civic bathing beach for Gary—and Mt. Baldy, the park's biggest dune, next door to Michigan City. The Beverly Shores beach had all of two people on it. This time, by contrast, Beverly Shores was packed. The parking lot the Park Service had built a quarter mile back from the end of Kemil Road was full. So was the much smaller lot on the little promontory that interrupts the beach in its middle: we found a place in this latter lot only because someone was backing out as we got there. The promontory had a new picnic shelter, with wooden stairways leading down to the shore. A set of interpretive signs explained the value of keeping Great Lakes shorelines natural. They didn't explain that the promontory itself was artificial—as was, in fact, the beach. The promontory owes its existence to riprap, big boulders piled up against its shoreward side as though a deranged giant had been playing with pebbles; the riprap absorbs the force of storm-driven waves and so keeps the land behind it—which contains, along with the picnic shelter, the National Lakeshore's research headquarters—from washing away. The artifice of the beach was revealed by a line roughly halfway from bluff to waveline where the sand abruptly changed color and texture. As with the rest of the beaches in the park, the Beverly Shores beach is maintained by artificial shore drift. Sand is brought in from elsewhere on the Lakes—often from dredged harbors—and dumped downcurrent from the groin complex at Michigan City. Enhancing beach nutrition in this manner is not new: Robert Fudge had shown me where they were doing it on an experimental basis in 1983. The coarser, darker sand was washing rapidly away, and Fudge commented that he didn't think it would be a problem for very long. Neither of us then could foresee my return fifteen years later, when beach augmentation would still be practiced and it would be Fudge that would be gone.

We drove west to a small parking lot adjacent to the gated community of Dune Acres, left the car and the crowds, and walked a mile in the company of humming mosquitoes and ancient forest-covered dunes into the heart of Cowles Bog. The bog is named for Henry Chandler Cowles, a University of Chicago biologist from the early years of the twentieth century who is usually credited with inventing the science of ecology. The Indiana Dunes were his laboratory. Walking transects back from the Lake across successive ridges of sand, Cowles noted that plant communities increased in both complexity and stability as you got further from the water. Out of that observation grew the principle of plant succession, one of the cornerstones of modern ecological thought. The trail into the bog followed an old raised roadbed for half a mile along a shallow, shrub-overgrown pond that flirted with being open water. A forested hillside rose beside the trail on the right. Cowles had been among the first to recognize that these steep little sand hills were actually old dunes, immobilized and stabilized by the trees and grasses that grew upon them.

Back in the company of numerous vacationers, we climbed the trail up Mt. Baldy. A high, tenuous overcast had begun to develop, but the light-colored sand of the giant dune was still bright enough to nearly boil the eyeballs. With big sliding strides like those of children, loose sand spraying everywhere, we descended lakeward. Eighteen-inch breakers swelled past laughing, bikini-clad teenagers to boom on the beach. Accents of gulls posed themselves against the sweeping form of the dune as if waiting for Andrew Wyeth.

Atop Baldy once more, preparing to descend to the car, we noticed activity at the base of the big dune to the east. Trucks moved back and forth in the shadow of the NIPSCo plant's cooling tower, dumping sand; a front loader appeared to be spreading the sand into the water. "Artificial shore drift," Melody guessed, and she was probably right; all that this massive collection of heavy equipment seemed to be doing was simply supplying sand to the longshore currents, so they could do the work that longshore currents are supposed to do. If that was the case, however, it certainly underscored the need to find a different solution to the problem. As fossil fuels become scarce in the next century—and all serious analysts expect them to—how long is it going to remain practical to import massive amounts of sand twenty diesel-driven yards at a time? Especially given the fact that nature, if allowed to, will provide far better service for free.

Above the tiny trucks at its base the great dune loomed impassively, an elephant contemplating the odd activities of some of its fleas—an analogy that would be more comforting if the fleas didn't carry a disease called avarice that is potent enough to kill the elephant and cart away its bones.

East of Michigan City we followed U.S. 12 into Michigan. Where 12 turned inland at New Buffalo we veered off onto Michigan 11, the Red Arrow Highway, up the west coast of the Great Lakes State. It would have been up the coast, that is, had it not been for the houses between the road and the water, nestled into woods or climbing over dunes, often obtrusive, always present. Many appeared to be new. Early in the trip, driving east toward the Lakes, I had expected the zebra mussel and the rise and fall of RAPs to dominate my conversations with Great Lakes residents. These things had been mentioned, of course: but nearly everyone with whom I had spoken so far had mentioned suburban sprawl as the number one problem facing the Lakes today. I was beginning to see why. Waukegan Harbor was mostly cleaned up: the waterways of South Chicago were on their way, and Green Bay was at least not getting any worse. We had yet to see a zebra mussel. Sprawl had been with us all the way from Door County.

We stopped briefly at a bluffside park in the lovely old town of St. Joseph, then pulled off the road again a mile or so later and went down to the edge of the Paw Paw River to watch a group of six tall-masted sailboats pass through a raised drawbridge. On the far side of the drawbridge, in Benton Harbor, we wandered little streets looking for the mouth of the Paw Paw. Ancient brick pavement led finally to a small green park called Tiscornia where the river blended quietly into the Lake. Late afternoon had turned the light golden beneath a high layer of cirrocumulus gathering in the distant west, over the big water.

The coast highway dead-ended onto Interstate 196 in a tiny town called Lake Michigan Beach. The map showed gray roads continuing north, near the water. We ignored them and took to the freeway. We were headed for Holland, home of tulips, windmills, and harsh Michigan winters, where we hoped to rendezvous before dark with a person I could easily have spoken to without leaving home—one of my coworkers from the Ashland Public Library, back in distant Oregon.

Amy Blossom grew up in Chicago, in a comfortably middle-class family with a lot of kids. She went to library school and moved west, first to Colorado, then to Oregon. There was a marriage, two daughters, a divorce, a second marriage. A brother who died young, of AIDS, in San Francisco. Through it all, Holland has remained a constant. The Blossoms have owned a lakeside cottage there, in a small vacation-home enclave called Castle Park, for many years. Amy has been returning to it each summer for as long as she can remember.

The route to Castle Park began as a nondescript alley between a couple of fast-food joints on Holland's main drag. A block away it right-angled onto a narrow city street, then right-angled again a couple of blocks later and became a little rural two-lane road headed due west. Six miles later it curved down to the left around an old tree-covered dune and ended against the small building that serves as Castle Park's post office. The Blossom cottage was just to the right atop a small hill, surrounded by trees and periwinkle, and Amy was on the back stoop to greet us. Dusk had fallen, and there was a sound of surf.

Inside the "cottage"—actually a hundred-year-old three-story frame house with a big sleeping porch—the air was bright with activity. Most of the large Blossom clan had come together at Castle Park, as they try to do each summer. Children of various ages ran through the house. Amy's husband, Brad Galusha, materialized at my elbow and handed me a can of beer. On the front porch Amy's daughter Ellie, age eleven, was the center of a small knot of laughing children that included her stepbrother Gallagher, age twelve. Ellie and Gallagher associate me with Amy's work. They seemed a little surprised to see me in Michigan.

We walked around the grounds in the gathering dark. Castle Park owes its name to a real, though small, castle, built years ago by a wealthy family as a vacation retreat, maintained today by the Castle Park community association as a town center. It contains conference rooms, lounges, a dining hall, and a small library. Upslope to the south an amphitheater nestles against a dune; community members present plays to each other there each summer. Near the beach stands a small youth center, complete with a professional recreation director. The community association pays the recreation director's salary, too.

At the center of the community, fronting the castle and surrounded by woods within which houses hide, there is a large open space; and in the middle of this greensward we encountered Amy's older daughter Meagan, thirteen, and stepdaughter Casey, fourteen, wandering hand in hand with a small person who appeared to be about three. "We hadn't been here an hour," Amy said, "when somebody called and asked if Meagan could baby-sit. People just seem to know when everybody comes and goes here." We passed others out walking in the dusk, or working in their yards, or just sitting, looking out at the woods or over the darkening Lake. All of them greeted Amy by name. It was only Brad's second visit to Castle Park, but most people seemed to know just about everything about him, too.

Written down, this description sounds busybody-ish and faintly oppressive. Experienced, it was precisely the opposite. There was a sense of familiarity and calm, a sort of F. Scott Fitzgerald air of relaxed elegance, as though Castle Park had been transported, intact, from the early-twentieth-century Hamptons. I half expected Bing Crosby and Grace Kelly to emerge from the castle in tennis whites. "This is a very safe place for children," Amy observed. "Everybody knows them all, and everyone looks after them. They can run around unsupervised in a way that would be unthinkable in town." It seemed a very safe place for adults, too. I could easily understand why Amy moved heaven and earth to see that she was always able to come back here for a couple of weeks each summer.

It is, of course, precisely this air of established community—this cross between an extended family and a summer camp for the well-to-do—that others hope to capture by building housing developments on the lakeshore. Like most imitations, they are rarely successful. Houses can be built wholesale among the dunes: communities cannot. "A town is saved," wrote Henry David Thoreau, "not more by the righteous men in it than by the woods and swamps that surround it." That goes for building towns as well as saving them. Castle Park became Castle Park when it was alone in the wet wild on the shore of a huge and lonely freshwater sea. Setting down a new group of big dwellings cheek by jowl to the next new group cannot hope to accomplish the same thing. There is a thing called the Fallacy of Composition that is very much in evidence around the Lakes these days. Simply put, it is this: things you want add up to things you don't want. Homes are built by the lakeshore to enjoy the Lake, but every home built there makes the Lake less enjoyable.

And thus Castle Park is threatened by all the communities that wish to be Castle Park. This threat is not idle, nor is it abstract. It is visible at the base of the bluff, where what used to be a broad beach has been reduced to a narrow ribbon of sand barely wide enough to walk along. Old-timers in Castle Park can pinpoint, with great accuracy, when the destruction of their beach began. It began when the Corps of Engineers built a groin to protect the newer community that had gone in next door, up the shore.

It doesn't seem fair that things should be this way, and of course it isn't. Life is not fair. It is not fair to Castle Park residents that newer communities shoulder them on the lakeshore and steal their beaches; it would not be fair to others who wish to live by the shore if coastal homebuilding had stopped with Castle Park. That, however, is the tradeoff we are stuck with. It is a tradeoff that finds echoes throughout the Great Lakes Basin. Twenty-seven million people simply cannot live as though each is the only one who lives beside the Lakes—or who loves them.

And so the dusk descending at Castle Park as we strolled the grounds seemed more than night approaching: it seemed the end of an era coming down. I didn't know how long the community could hold out against it. How many more years would the Blossoms and their neighbors gather in the shadow of the castle to show their plays and share their children's joys and watch the sun sink like a bright stone into Lake Michigan?

All too soon the evening came to a close. We drove back to our motel beneath a starless and moonless dark that held the close smell of impending rain.

IV
THE MICHIGAN MITTEN

It was drizzling in Holland when we got up the next morning: the town looked gray and dismal, as if someone had opened a tap and drained the color out of it. We tried to play the Breakfast Game along the bypass east of town and had to be satisfied with a chain roadhouse at the entrance to a closed mall. Up the road a piece, the visitor center at P. J. Hoffmaster State Park was shut up tight when we arrived shortly after 9:30 A.M. The visitor center had the park's restrooms in it. It began to look as though the Michigan Mitten had it in for us.

"The Mitten" is what Michiganders call the lower half of their state—the part officially known as the Lower Peninsula. One look at a map will tell you why. The Lower Peninsula looks like an immense mittened hand. Lake Michigan and Lake Huron sweep toward each other around the peninsula, meeting in the Straits of Mackinac at the tip of the index finger; Saginaw Bay separates the Thumb from the palm. It is said that you can tell Michigan natives by asking where in the state they live. If they point to a spot somewhere on one of their hands, they were probably born here.

The northern part of the Mitten is mostly moraines, masses of them, left by the continental glacier and tumbling every which way across the land. It is a hardscrabble country. Once it was covered with big white pines; then the loggers came through, more than a hundred years ago, and for a while it was bare. Farmers arrived, scratched the poor soils for a few years or a few decades, and left, defeated. Now there are pines again. Small, clear lakes drain down little rivers into the freshwater sea. The place grows natural beauty far better than it ever grew crops. The cuff of the Mitten, a broad band stretching from Kalamazoo to Detroit, holds most of Michigan's population; but almost all of them dream of living up here.

And, increasingly, they do. Shorelines sprout vacation cottages and retirement homes; towns are platted among the pines that crown the moraines. It is not a new phenomenon—Castle Park demonstrates that—but it has accelerated in recent years, and it has changed character. The houses are bigger and they sit closer together, on smaller lots. The infrastructure—paved streets, sewers, phone and power lines—strongly resembles that of the urban areas the residents of the houses presumably came up here to escape. Bob Beltran and Kent Fuller had talked at length about this phenomenon back in Chicago. Now I was seeing it.

The problem is most acute along the west coast of the state, the part of the Mitten occupied by the little finger. Here the glaciers left behind massive amounts of sand, raw material which the wind and the Lakes have shaped into the most extensive set of freshwater dunes on the planet. Dunes stretch in a nearly unbroken line from Indiana to beyond the Straits of Mackinac. The successional pattern identified by Henry Chandler Cowles in the Indiana Dunes holds for the rest of the dunes belt as well: barren foredunes, mid-dunes pinned down by marram grass, tree-anchored and stabilized sand hills further back. Toward the north, the dunes often perch on bluffs of glacial drift with small beaches at their bases. The culmination of this bluff-and-perched-dune pattern comes at Sleeping Bear, where the dunes sit atop a steep, sandy slope more than four hundred feet high.

The line of coastal dunes appears unbroken on a map—and I have just described it so—but this is a bit misleading. There is a hidden story in the topography here that only shows up when you study the map closely, or when you drive through the countryside. The predecessor of Lake Michigan, giant Lake Nipissing—which also encompassed Lake Huron and Lake Superior—had a deeply indented coastline along what is now the Mitten's smooth western shore, with great bays fingering back among the moraines like fjords. Nearly all these embayments have become isolated by falling Lake levels and dune building, and are now separate lakes. It is this rhythm of small lake and large, surrounded by dune and strand, that has lured homebuilders here. Waterfronts and beaches always make the most attractive homesites. There is a greater concentration of waterfronts and beaches in the western Lower Peninsula than anywhere else in the Midwest—perhaps anywhere else in the world. So the dunes sprout houses, and houses, and more houses, until there are more houses than dunes and one wonders why anyone would bother anymore.

There are breaks in this pattern, of course—places where public ownership has put small bits of the best of the coast under lock and key. Anchored by National Lakeshores at the two ends—Indiana Dunes on the south, Sleeping Bear Dunes on the north—numerous local parks, several state parks, and one National Forest wilderness area keep as much of the dunes belt as they can inviolate. It is a noble effort, but a losing one. Three things constantly work against it. The first is rising use levels. The second is falling revenues. And the third is the belief that, in order to protect a natural landscape, you must somehow isolate it from all human-caused effects—a belief that seems to become more and more strident the more it is squeezed between rising visitor levels and declining dollars.

The sign on the door at P. J. Hoffmaster's visitor center said it would open at 10 A.M. We turned away from the locked building with its inside restrooms and headed for the nearby dune climb. As we passed the parking lot, two large tour buses pulled up and disgorged approximately fifty passengers each onto the tarmac. Were *they* going to be surprised.

The dune climb was a series of stairways separated by short lengths of path, lovely and solitary—not a single person from either tour bus followed us. Near the top the path split, one branch leading up and left to the crown of the dune, the other traversing right and slightly down to a viewing platform at the edge of a large blowout. The upper route was barricaded. There was evidently fear on the part of the State Parks Division that someone might fall off the deteriorating platform on top of the dune, and break something, and sue. It was a fear that was probably all too well founded. Lawsuits and the threat of lawsuits are fast turning us into a nation of wimps. We stayed on the legal side of the barricade. I guess I am as wimpy as the rest.

Back at the bottom of the trail the visitor center was finally open. Aside from the restrooms, it appeared primarily to contain a gift shop. I asked the girl behind the counter about the closed platform at the top of the dune climb and got a blank look: she was not even aware that the closure existed, let alone why it had been done. She was a nice person but she seemed overwhelmed, the inevitable position of public employees pinched between simultaneous demands for escalating services and lower taxes, and if she didn't know enough of what was going on in the park to keep visitors properly informed, the visitors—if they lived and voted in Michigan—had no one to blame but themselves.

Outside, the drizzle had ended and the sun was beginning to burn through the dissolving overcast. By the time we reached Ludington the clouds were gone completely and the sky over the Lake had turned a bright, piercing blue. We bought food at a deli in a large new supermarket, part of a commercial strip that hadn't existed at all fifteen years ago, and took it to Stearns Park, on the waterfront. There were tables in the shade, and breakers rolling onto a large, flat, uncrowded beach. Ludington is the eastern terminus of the last remaining ferry across Lake Michigan: the western terminus is at Manitowac, Wisconsin, which we had passed through a week before. Someday I plan to take that ferry across the Lake, but someday keeps getting postponed while the boat and I both age, and it is quite likely, as with many dreams, that it will never actually arrive. We put away the lunch, got back in the car, and hightailed it up U.S. 31 toward Frankfort, where I had a date to talk to a man about a harbor.

In 1983, in Frankfort, I had interviewed a Sleeping Bear Dunes resource manager named Max Holden. Much of our conversation had revolved around a demand from the state of Michigan that the National Lakeshore build a harbor of refuge—a protected anchorage recreational boaters can retreat to when storms arise—somewhere along its section of the coast. It is state policy that such harbors be no more than fifteen miles apart. The National Lakeshore is forty miles long, and there was no such harbor its entire length. The Park Service wanted to keep it that way. "We wouldn't mind just building a dock," Max had told me, "but that's not what they want. No one is hearing us. We're talking about a dock and people are hearing 'harbor.' It may happen like a juggernaut. We're saying we don't want the responsibility to provide overnight slips—we don't want to sell bait. But nobody's listening."

Unlike my contacts at Indiana Dunes, Max was still at Sleeping Bear. We had established a probable date for my arrival, but not a probable time. It therefore seemed important to get to Frankfort quickly, so I could catch him before he gave up and left for the day. We did that, pulling into town shortly after 4 P.M. But there was no Lakeshore headquarters. I found the building where I remembered meeting Max: it was empty. The only sign in town that seemed to have anything to do with the Lakeshore pointed up State Highway 22, toward Empire.

We stopped at the Chamber of Commerce and spoke to the woman at the information desk. Oh, no, she said, the Lakeshore headquarters wasn't in Frankfort. She didn't recall that it had ever been in Frankfort. Like the sign, she pointed up Highway 22. Go to Empire. There's a big building just off the road there, to the right. That's the visitor center. Ask there.

It was disorienting, and I couldn't rid myself of a nagging fear that we would have to come back to Frankfort, after all, to find Max; but there was no point in disputing the Chamber's information person and no time for dispute anyway. We beat it up the road toward Empire, racing the hands of the clock. Empire was a tiny little shut-up town with a combination gas station and grocery store and a couple of restaurants and, sure enough, a large new Park Service–style chateau off to the right. We pulled into the parking lot. It was 4:35, and when we went inside, Max was there.

WE FOUND A CONFERENCE ROOM. "WHEN
did you move into this building?" I asked. "Seems like I saw you at
Frankfort, before."

Max thought for a moment. "It's been ten years, I think. Eight for
sure, and probably more."

"I was looking for your headquarters down in Frankfort."

"Oh, dear." He laughed. Max is short, robust, and soft-spoken, and
his gray Park Service uniform hangs uncomfortably on him; he looks
as though he would be much happier in a flannel shirt with a canoe
paddle in his hand. His hair, which he wears shoulder-length, had
grayed considerably. Fifteen years will do that to you.

"I asked at the Chamber of Commerce," I said, "and they said, 'No,
it's up in Empire.' And I had to do a double take. Because—"

"Well, that was always our long-range plan, to move into the park,"
he explained. "And that building in Frankfort was available until
we could make plans." That was a relief—I had remembered right
after all. "This is a leased building. It went out for bids. The General
Services Administration asked somebody to build a building that
would meet our needs for a visitor center and offices, and they got
something like a twenty-year lease to do that. It's worked out real
well. Empire's right in the middle of the park, and it's a really nice
building for us."

"The last time we talked, fifteen years ago, one of the things that
was hanging over your head was a requirement to build a harbor of
refuge. What's the current history on that?"

"Well, the Park Service said 'No' at that time, but it's come up
at least two times since. One place that was suggested was at Glen
Haven, where the cannery and the old dock were. And our planner
said it was feasible. You could build a harbor there, it was just a matter
of money." He chuckled. "Well, maybe *anything* is feasible, with the
money. But by the time they built a breakwater, and parking lot, and

everything, it would be enormously expensive. And although the Park Service said it was *feasible,* and perhaps the *only* place it was feasible in the Lakeshore, the neighbors just rebelled at that. They really scolded the Park Service. They said, 'It's your obligation to protect these natural beaches, just like you said you were going to.' So it fizzled. It was just not the will of the people to do it.

"But it came up again at the mouth of the Platte River. We've got a big sports fishery down there, with the coho salmon coming up every fall, and it's really popular. For years we've been dredging a channel, starting after Labor Day, so that fishing boats can launch upriver from the mouth—"

"On the Platte," I repeated, not wanting to believe I had heard right. In 1983 the Platte, at the south end of the park, had possessed one of the loveliest river mouths I have ever seen. I had watched a sunset there that was quite possibly *the* loveliest I have ever seen.

Sunset from the mouth of the Platte River, Sleeping Bear Dunes National Lakeshore, Michigan

"On the Platte," said Max. "And—"

"How many years has this been going on? Can I ask?"

"Oh, well—I would say twenty years or more, now. And the state was doing it before we did." I relaxed. They were doing it cleanly enough I hadn't noticed. "The sports fishing coalition and some of the business people formed a commission to lean on the Park Service to build a harbor there," Max continued, "and again we said no. We would continue to dredge—that's really contrary to our policy, to do an unnatural thing like dredging, but it seemed to be the least environmental impact. There's not a lot of invertebrates or fish spawning right at the mouth, and the dredge spoils are washed away, or leveled at least, by the winter storms. So we said we would do that, rather than build a harbor. That's our position still."

"OK." I felt somewhat mollified.

"In the meantime," Max went on, "the piping plover, an endangered species, has started to nest in that location, too."

I was suddenly unmollified again. "On the Platte River?"

"On the Platte River."

"On those bars down there?"

"Yes. So that's all the more reason not to do it. But the main reason is, it's probably the last natural river mouth on Lake Michigan."

"It's a very, very beautiful spot," I said.

He nodded. "So we're still holding the line on the harbor. And I expect we will continue to do that."

"How's the state of Michigan reacting?"

"Well, the Waterways Commission leaned on us very early, when the park was established, but some of those people who were so adamant about it have retired and gone out. Of course, the state would still like to have a harbor of refuge."

"I think there's a state law requiring them every so often around the Lake," I said.

"Well, I think it's a policy, not a law," Max said. "It's the Waterways Commission's policy to have a harbor of refuge every fifteen miles. And it would be ideal for the Waterways Commission to use federal money to help them meet their goal. But it's not *our* goal."

"OK," I said. "It doesn't appear that the harbor of refuge situation has changed significantly. What have you observed that *has* changed?"

He didn't seem to have to think at all. "Things have just happened so quickly with the introduction of exotic species," he said. "And we don't know the full impact of that yet. The lamprey eel has been with us for a hundred years or more, ever since the Welland Canal was finished. But now we get the river ruffe, and the zebra mussel, and the spiny water flea—just one thing after another, it seems like. All of these things are so new, but so quickly developing. We're in a position now where we would like to do something, but it's almost too late. You go down to the shoreline after a storm, and every one of these little rocks already has zebra mussels on it. You see them washed up on the beach, and you know that's just a miniscule part of the population. At the power plant at Traverse City, where they're discharging warm water from the cooling of the power plant, the rocks are just covered with them. We have them on our dock at South Manitou Island. There's some controversy about how much impact they are, or are not, having. They're probably food for ducks in some places. But they're clogging water intakes. I don't think we really know what's going to happen."

"No," I agreed, "they're too new to really have a clue. It's one of the big unknowns. But—so you think exotic species are the main change?"

"I think it's probably the most significant change in the last ten years."

"Interesting," I observed. "I thought as I came into this project that that would probably be what most people said. But you're the first person who *has* said it. Mostly what I'm hearing from people is that the big issue is suburban sprawl."

"Yeah, well," he said, "that certainly happens. It's been an ongoing thing here at the edge of the park. And along with it comes noise, and light pollution, and things that sometimes you don't think about. One of the things that the Park Service is going to be taking some good and bad publicity about, I guess, is our objection to jetskis. Jetskis is a brand name, so I guess I shouldn't say that. Personal watercraft."

"Yeah," I said, "that's the term that's usually used."

"Yeah. The Park Service now, on the national level, is seeking to ban jetskis from the Park Service areas. The regulation hasn't been written yet, but they're working on it, and it's been announced to the press. We've had some local objections already in the paper."

"Wouldn't surprise me at all," I said dryly.

Max laughed. "The local sheriff has said it's discrimination to ban that kind of watercraft and not ban other boats. And that may be a point, but they are a category of boats that really stands *out* as being objectionable."

"Would that just be the inland waters?" I asked. "Or are they a problem on the main Lake?"

"Well, it would be both," Max answered. "We would seek to ban them on the inland waters too, but they're a nuisance over on South Manitou Island, and a lot of the shoreline here."

"They can go to South Manitou?"

"They *do* go there."

"They do."

"Often. On a nice day there'll be half a dozen or so in the harbor over there, just buzzing around."

"Wonderful. Do people put them on boats and take them over there, or do they actually—"

"No, they run 'em across."

"People run 'em across." I was having a hard time believing such a small, sinkable craft that far out on such a big body of water.

"Yeah, they're fearless," Max said. "A friend of mine has a kayak, and we were out weekend before last—just a quiet evening out in front of Empire—and there were two jetskis. It was Anchor Days, so there was a big crowd down on the beach, and boy, the noise level—they do it to show off, you know. They could easily have gone down the beach, just as we did with the kayak, to get away from the crowd—nah, they just went back and forth, back and forth—"

"Yeah," I said. "That's the point."

"—and did all the little tricks. There was a recent article in the Traverse City paper about evidence that jetskis put more unburned fuel in the water than any other kind of watercraft. So there's that factor. And people talk to you on the street and say, 'Boy, I hope you *do* ban 'em.' We get some support."

"Have you seen any organized opposition?"

He sighed. "No, but there will be. The people that interviewed the sheriff also asked the local salespeople, and their response was, 'It's just a few bad apples, and the industry is working to tune down the noise, and the pollution.' Trying to bring out the unfairness of the Park Service to ban 'em. But it'll probably happen by the end of this year."

"I would presume you have wind surfers, as well."

"Yes."

"And they would probably not cause the same kind of problem."

"No, not at all. I guess we feel that way about hang gliders, too. We allow radio-controlled gliders—models—we don't allow—"

"Hang gliders?"

"The engine ones," Max explained. "We allow hang gliders, but—"

"Nonmotorized," Melody supplied. "Do you already allow motorized boats of other classes?"

"Yeah," said Max. "Oh, yeah."

"So the problem is," she said, "why with one motor, and not with another?"

"Yeah. The sheriff pointed that out."

We talked some more about the piping plover. "That's been a real success story for us," Max stated. "We actively go out in May and look up and down the beaches to find them, and we put little fences over them to keep predators out. And down at the mouth of the Platte River we fenced off a generous portion of the beach so that the people wouldn't use it. It was just a psychological fence, we put some signs and baling twine, but with the biological technicians—they're down there monitoring the birds to see when they hatch, but they also greet almost everybody that comes there and give them the story about the piping plover. We have a spotting scope set up at a distance, so they can see 'em. And you know, most people have never *heard* of a piping plover, let alone seen one.

"There was some opposition, of course. You know, 'It's human beings that are the endangered species' "—he laughed—" 'and we've been walking our dog here for twenty years, and those birds have always been there, and we've never bothered 'em'—things like that. But we've had really good compliance, because we've had people in uniform right down there all the time. So it was a good year for that. But we've not had good luck with our eagles. When I talked with you before, we hadn't seen eagles nesting in the park for twenty years previous to that. Well, about 1993 we had an eagle start a nest on North Manitou Island. They had four years of successful nesting, and they raised two chicks each year. And then last year the nest failed. People I talked with in the Fish and Wildlife Service thought it might have something to do with the real severe, lingering winter that we had that year. This year there's another nest—not in the same place, but we think it's the same pair. They have had one chick. And then we had a nest start on the mainland last year, in a swamp up by Shell Lake, and they raised one eaglet. This year that family started a nest, and added to it, but I don't think they laid eggs. They just abandoned the project."

"We've seen a lot of purple loosestrife," I commented.

Max nodded. "We just introduced the Asian beetle in the park in one location. All the purple loosestrife at South Bar Lake—Michigan State University released a couple of thousand of them in there. And we just released five hundred in a little plot that we have. We don't think that's the way to go, but we're going to try it, to—"

"—see whether in fifteen years it's the Asian beetle that we're worried about," Melody finished for him, and we all laughed.

"Well, the Park Service is really conservative about it," Max said. "Other people were doing it right along, and we've been reading about it, and—we were just slow. But Michigan State University people were doing it around the park, and they asked us to do it. We talked to our people, and they were cautious, but they said, 'Frankly, they're releasing them all around you anyway, and there's nothing you can do about it.'"

The visitor center was getting ready to close. Max walked us to the front door.

"I'm glad you were still here after all these years," I told him.

He nodded. "You know," he said, "after my divorce I applied for a job at Pinnacles, in California. My son is out there, and I thought it would be a good opportunity. But I didn't get it." He smiled. "I found out I was relieved. This is where I really want to be."

We drove the Pierce Stocking Scenic Drive in the late-afternoon light, winding through the woods and dunes to the edge of Sleeping Bear Bluff. Pierce Stocking was a retired logger who lived in the area and wanted people to see the beauty of the dunes. He built the drive and operated it as a toll road from 1967 until his death in 1976. His heirs sold it to the Park Service, and for a time access was free. Now it is a toll road again: a Park Service fee station has been set up near its beginning. Max had mentioned some recent changes to the route, but I didn't spot them; however, the overlooks I recalled as simple trails through the sand had turned to wooden walkways and viewing platforms. At the bluff an unofficial trail led over the edge and down—straight down, 423 steep, sandy feet to the edge of the water. Many people were taking it. We stood on a platform jutting out from the rim and watched them bound down that long slope and then struggle back up a few sliding steps at a time, and it wasn't even tempting. The Lake stretched to the blue horizon, flecked with whitecaps. Afterward we picked up sandwiches and fudge at a tiny deli in Empire's combination gas station/grocery, set up our tent in the park campground, and headed for the mouth of the Platte. The light was warm and clear. We stopped to eat the sandwiches at a boat ramp a mile or so upriver from the Lake—a single table, a small wetland, and lovely reflections in the quiet water—and then continued to the mouth, where we spent over an hour watching the sunset and trying to capture its elusive light on film. To the north across Platte Bay the bluff beetled up, its white face turned yellow by the long westering sun. The mouth of the Platte is oddly bent: scarcely a dozen feet from the Lake the river turns sideways and runs parallel to the water's edge for more than a quarter of a mile behind a low, skinny sandbank before finally breaching through to the Lake over the shallow bar Max had told us the Park Service dredges out each fall. We waded across the bar to the narrow finger of outer beach, where breakers curled in from the horizon to fall upon the sand with a sound like that of the sea. The sky and the backlit breakers glowed orange. It was an evening much like the one I had spent there with Larry Chitwood fifteen years before, only that one had made each of us wish our wives were present and this time mine was here. All too soon the show ended and the dark came down. We went back to the campground and our tent, and I won't tell you what we did next.

I do need to say something about camping in the Lakeshore. In 1983, Larry and I had arrived in late afternoon on a weekday and had no trouble claiming a site. This year we were also there on a weekday, but I called two months ahead for reservations and was told the site they gave me was the last one available. Larry and I shared our Platte River sunset with perhaps ten other people; I made no count this year, but the number was probably someplace between fifty and one hundred. I do not begrudge anyone else their campsite or their place in the sunset, but there is no question that today's larger numbers diminish the experience. The Fallacy of Composition sits heavily over Sleeping Bear National Lakeshore, and it is unlikely to ever go away.

THE THING YOU FIRST NOTICE ABOUT THE
Leelanau Peninsula is that it goes up and down a lot. At least eight
big north-south moraines lie parallel to each other here within a space
of just thirty-five miles. The first of this set of long linear hills runs
up the east side of Grand Traverse Bay; the next one west forms
the backbone of the long, skinny Old Mission Peninsula, separating
the bay's east and west arms. The remaining six march rhythmically
westward across the Leelanau to Lake Michigan. Taken together, this
series of parallel ridges forms one of the most striking landforms left
behind when the glaciers abandoned the Lower Peninsula.

State Route 72 crosses this corrugated land at right angles, rising
and falling like a long-period roller coaster as it cuts across the base of
the Leelanau from Sleeping Bear to Traverse City. We hit it in the early
morning, heading toward Traverse after an unsuccessful attempt to
find breakfast in closed-up Empire. The west sides of the moraines
were still in shadow and the east sides were in sun, so we traveled
through a sequence of dark and light as well as up and down. Farmers
on tractors stroked the warm skins of their fields. The sun played like
a kitten with the bright toy of the roadside dew.

Traverse City spread like a miniature Naples around the base of
its blue namesake bay. Highway 72 came down the last moraine,
hit the bay, and turned southeast along the waterfront. To the right,
residences, then businesses; to the left, a narrow ribbon of park going
down to the water. The road ran out on a skinny east/west finger of
land separating the Boardman River from the bay. We had still seen
no open restaurants. I was about to give up on the Breakfast Game
and hit a fast-food place—I had actually spotted the golden arches,
several blocks ahead—when a sign painted on the side of a building
on the far side of the little river caught my eye. It read "Breakfasts—
Lunches—Catering." We took the next bridge and doubled back

down Front Street to the building's entrance. An almost invisibly small sign identified it as the Left Bank Café.

The place was tiny, with seating for no more than twelve or fifteen people. The single restroom doubled as a mop closet; to use it I had to pass directly through the small kitchen, nearly rubbing elbows with the cook. The wooden tables were bright and clean and covered with checkered cloths. The menu seemed the size of a small bedsheet, and everything we sampled from it was superb. Scrambled eggs, moist and golden and piping hot. Thick slices of toasted homemade bread, slathered with what appeared to be locally made preserves. Oven-roasted potatoes smothered in butter and basil. Omelettes of all possible description. The coffee was dark and rich and so fresh it should have been slapped. On the Breakfast Game scale of one to ten, the Left Bank Café ranked somewhere above eleven. I have never had a breakfast like that anywhere else on the continent.

After dallying far too long on the Left Bank we strolled out into the midmorning sun to find the parking meter run down but the car unticketed. The gods were smiling. We found our way back to Michigan 72. In a few short blocks we turned left, onto Route 37. Soon we were riding the high backbone of the Old Mission Peninsula, covered with cherry orchards like the Door but, except for the part down by Traverse City where the usual sprawl was starting, far less developed. From the crest of the narrow peninsula water was visible on both sides. We meandered north as far as Mapleton, then turned around and meandered south again. Back at the base of the peninsula, in Bryant Park, an outdoor art lesson was proceeding: a dozen or so middle-aged students clustered around an easel and watched their teacher limn the blue bay onto blank canvas. The temperature was shirtsleeve-wonderful. The guy who wrote that song that locates heaven in West Virginia had obviously never been to Traverse City.

U.S. 31 ran up the east side of Grand Traverse Bay to Charlevoix, where we found all the tourists who had been blessedly absent from Traverse City. The narrow isthmus separating the big Lake from Lake Charlevoix was full of buildings and people. We escaped down Michigan 66 along the southwest side of the lake. Like most of the lakes that line the west coast of the Lower Peninsula, Lake Charlevoix is an isolated embayment, a former twin of Grand Traverse Bay, divorced from the big Lake by drifting sand and declining water levels. Halfway up its twenty-mile length, Charlevoix splits. The main body of the lake goes off to the east, toward Ernest Hemingway's boyhood Horton Bay; the much narrower South Arm heads due south. At its southern tip it receives the waters of the Jordan River. The Jordan is a vein in the ring finger of the Mitten, draining much of the northwest corner of the Lower Peninsula. The lower end of its valley is a narrow, forested V. The upper end is old lakebed, flat and dark and full of rich, loamy soil—perhaps the only truly good cropland in the northern Lower Peninsula.

But there is a worm in the Jordan Valley's apple of Eden. Under the valley rests the Antrim shale, a black, friable rock formed of marine sediments laid down beneath the shallow sea that covered Michigan at the close of the Devonian, approximately 400 million years ago. Shale is compressed mud, and its color indicates the mud's organic content: the darker the color, the more organic material. High organic content in the original mud translates, in most cases, to high petroleum content today. In the Antrim shale, the high petroleum content has expressed itself as large quantities of natural gas—a gigantic earth fart, petrified in stone. Natural gas migrates upward between the tilted layers of the shale, collecting in pockets. There are thousands of these pockets of gas under the Jordan Valley and its surrounding hills.

As late as the 1960s, the amount of gas here was thought to be trivial. In the 1970s, it was discovered that it wasn't. The rush began in earnest in 1989. Since then, more than six thousand wells have been drilled in the Antrim shale, making the region surrounding the Jordan Valley the most intensively developed gas field in the United States. The valley itself remains inviolate—or at least the lower part of it does. Under the terms of its management plan, the Jordan Valley Management Area—twenty-two thousand acres of largely state-owned, largely wild land—has been off-limits to oil and gas development since 1975.

Ever since the Antrim shale rush started, Midwestern environmentalists have been wondering how long the Jordan Valley prohibitions would hold. They received a partial answer in 1997. In November 1996 a businessman named Walter Zaremba, from the tiny town of Elmira in the upper, agricultural end of the Jordan Valley, announced plans to develop a gas well on forty acres of land inside the Jordan Valley Management Area which he had purchased on speculation two years before. Zaremba requested an exemption from the drilling moratorium, and a waiver of a Michigan state law requiring a minimum of eighty acres between gas wells.

At first it appeared he would succeed. Both the state's Department of Environmental Quality and its Department of Natural Resources seemed primed to roll over and acquiesce to development; the DEQ, acting without public hearings, actually issued the eighty-acre waiver. However, when this action initiated statewide protests—including a resolution condemning the proposed well from the county commissioners of Zaremba's own Antrim County—the agency backed off. In May 1997, with Governor John Engler himself citing the need to protect the natural qualities of the Jordan Valley, the state denied the drilling permit.

At East Jordan, astride the river where it flows into the South Arm of Lake Charlevoix, we picked up Michigan 32 and followed it eastward through the morainal hills that lie in the center of the great U the Jordan scrawls across the upper Lower Peninsula. The highway does not traverse the moratorium lands, so signs of gas development could be expected. We watched for wells. At first we saw none; then Melody said quietly, "There's one." In a small clearing in the woods beside the road a metal box four feet square sat on a concrete platform. A pressure-relief valve jutted from its top; a delivery pipe sprouted from its side and elbowed down into the ground. The whole thing was painted green, as if trying to hide against the foliage.

After that we knew what to look for, and we saw many. Some were painted green, like the first; others were white, blue, red, or yellow. Some sat at the edges of pastures or fields; others nestled among the trees. They were quiet and unobtrusive. If this were all there was to it, one could be excused for wondering what all the fuss was about.

But of course that isn't all there is to it. There is exploration and drilling, which is much noisier and more intrusive than these inconspicuous little roadside boxes. There are processing plants, and pipeline corridors, and access roads. In its ten years of intensive development—all since *The Late, Great Lakes* was published—the countryside over the Antrim shale has lost somewhere between 500,000 and 1,000,000 acres of land to gas development. One wonders where the voices which complain about the "lockup" of wilderness lands have gone. Why do they not protest against this lockup of a far more single-minded and exclusionary sort?

The worst thing about the damage here, though, is that it really doesn't have to happen at all. In 1980, a decade or so after the extent of the Antrim reserves had become fully known, an agreement was hammered out in the Pigeon River Valley a couple of dozen miles to the east of the Jordan. At stake was the Pigeon River State Forest, nearly 100,000 acres of marsh and stream and magnificent second-growth white pine. The agreement, put together by representatives from state government, local communities, environmental groups, and Shell Oil, kept drilling out of more than two-thirds of the forest and placed wells within the remaining one-third in a few tight clusters. That kept the pipeline corridors and access roads to a minimum. Slant drilling enabled Shell to reach deposits beneath the closed area. The oil company agreed to submit its development and operations plans to a citizens' review committee to make certain that its impact on other uses of the Pigeon River landscape was minimized as much as possible.

One would think, given the success of this approach—I am not aware of a single complaint, either from the oil company or from the Pigeon River's protectors—that it would be widely copied. Nothing like that has happened. Over the rest of the Antrim shale it has been business as usual, which means a full-throttle rush to development by industry and a squealing of hastily applied brakes by environmentalists. Why cannot both sides understand that good driving means neither screeching to a stop nor cramming the accelerator to the floor, but simply staying within the limit?

In Gaylord, east of the Jordan, Antrim shale money had spawned the usual strip developments and subdivisions. We swung onto Interstate 75 and headed south. Half an hour later, a few miles north of Grayling, we turned northeast onto Michigan 93. In a few minutes we were lining up behind two other cars at the entrance station of Hartwick Pines State Park.

Mindful of the dramatic increase in camping in the Midwest since 1983—and the lack of any concurrent increase in campground construction—I had attempted to reserve a campsite at Hartwick Pines before we left home. I was unsuccessful. Michigan requires those reserving campsites in its state parks to be destination travelers committed to staying at least two nights: we would only be there for one. I am not sure what Michigan hopes to accomplish with this policy, beyond annoying those who wish to see all of the state rather than glue themselves to a small part of it. To me it feels like discrimination—a declaration that people who can afford to plan a longer stay are somehow more worthy than those who can stay only one night. But that is a theoretical and, I have to admit, sniveling aside. The relevant fact here is that I was driving up to the entrance kiosk at Hartwick Pines at 3:30 in the afternoon on a Friday in high summer, heart in mouth and wallet in hand, seeking a one-night stand as desperately and eagerly as any short-run traveling salesman.

We didn't get it, of course, although it looked for a moment as though we might. The crisply uniformed young man at the kiosk window thought he had something left. The woman at the mapboard behind him shot that down pretty quickly, though. She had given the final site to a couple in an RV a few minutes before.

"So what do we do now?" I asked, aware that I sounded somewhat frantic.

"You might try Jones Lake," the young man replied. "It's about ten minutes from here. There's a state forest campground there that's pretty nice."

"Do you know if it's got sites available?"

"Haven't had a recent report."

"Can you call and find out?"

"There's nobody up there to call. It's pretty primitive."

We whipped a quick U-turn in front of the kiosk and headed for Jones Lake.

Jones Lake was a kettle—a hole in the blanket of glacial drift that spreads over vast portions of the palm of the Mitten. A giant block of ice had fallen from the glacier's face during its retreat, lasted long enough for a thick layer of soil and rocks to gather around it, and then melted away. The lake was in flat country, in a wood of mixed black oak and red pine. It was roughly a quarter-mile wide by half a mile long, with a marsh at one end, and its gently lapping waters were nearly invisibly clear. The primitive campground beside it had pit toilets and hand pumps and only a scattering of occupants. Late-afternoon light had begun to slant in, and the wood and the lake seemed to glow.

We set up the tent and wandered over to the water. A small lawn dotted with picnic tables sloped down to a little beach. Melody went back for her swimsuit while I sat on the grass and tried to write in my journal. It didn't work. A large, boisterous family was playing with a paddleboat in ten-foot-deep water a hundred feet or so off the beach. The family included numerous small children: none were wearing life jackets, and they didn't seem adequately watched. I put the journal away and watched the unwatched children while the long light grew longer and Melody swam a discreet distance away from the paddleboaters.

Nothing untoward happened, although one overheard comment bears repeating. Part of the group had returned to the beach, and several of the children were splashing about in the shallows. One small boy complained that the water was too cold.

"Pee in it," his father responded immediately.

Say what?

"Pee in it. It'll warm it up."

And I suppose it would, at that. Still, I profoundly hope the boy didn't follow his father's advice. Fish may indeed do it all the time, as the father pointed out, but as cold-blooded organisms the volume of waste they produce is much less, and its chemistry—which the lake's web of life has adapted to—is completely different from ours. And fish don't carry pathogens that humans can easily pick up. Peeing in a lake may seem a small thing to do, but it is only slightly less antisocial than carrying a gun to school.

"I'm glad I was around the corner," Melody said afterward, when I told her.

We returned to Hartwick Pines the next morning. I am sorry to say we were somewhat disappointed. Oh, the trees were there, all right, nearly fifty acres of them, tall and straight-boled and dark beneath a thick canopy of needles that let only a little light seep in from what was, outside the grove, a bright and glorious day. But the largest were barely four feet through, and when you come from the land of the Douglas fir and the redwood a four-foot-thick tree is pretty small potatoes. Rod Badger and I had observed this same problem fifteen years before in the Estivant Pines, a reserve near the tip of the Keweenaw Peninsula in upper Michigan, where we had hiked half an hour over corduroy trails through mosquito-filled swamps to reach a "giant white pine" that, as Rod put it, shouldn't be put out to pasture with a Douglas fir because it would get eaten alive. The trail through the Old Growth Grove at Hartwick Pines was asphalt, not corduroy, and there were only a few mosquitoes around. It was cool and pleasant and nice in there. But anyone who really wants to see big trees should come to Oregon.

The Hartwick Pines bring up another problem, too—one that bears more discussion than it gets. The park sprawls over more than nine thousand acres of the central Lower Peninsula. Only forty-nine acres of that are devoted to old growth. The forty-nine acres used to be eighty-five acres, but a storm came through in 1940 and knocked thirty-six acres flat. The remaining acreage is a tiny droplet of old-growth white pine in a great sea of second growth. It is extremely vulnerable. Another storm like the one in 1940 could do the whole thing in overnight. The pines are also endangered by visitor pressure: the asphalt trails—and the small but emphatic signs every few yards warning you to stay on them—were put in after soil compaction from human feet destroyed several of the largest trees and endangered others.

All of which leads to an inescapable truth. Hartwick Pines State Park is not a forest preserve: it is a museum. And as with all museums, the value of the exhibits depends at least as much on the curator's skill as it does on the objects placed on display.

I do not begrudge Michiganders their park. But if you are looking for a sample of the pre-logging Lower Peninsula, you will not find it here. The tens of thousands of square miles of white pine that covered the state before the loggers arrived were not vulnerable to the effects of a single storm, nor to the feet of tourists trying to capture a little of the elusive past along a 1.25-mile loop trail. They did not need to be herded into an enclosure and guarded: like Kipling's cat, they walked by their wild lone. Rapacious logging put an end to that. Preserving this infinitesimally small remnant will not bring it back.

At the end of the Big Cut, at the close of the nineteenth century, Michigan was denuded. Today, most of that bared landscape is covered with timber again, and much of this regenerated forest is approaching one hundred years of age. There are millions and millions of acres of it. It is second growth, but it is mature. If you want to see what Michigan was like before the loggers got here, go to these old "new" woods and look there. I must warn you, though, that you don't have very long. Second homes are devouring the second growth, and big timber is poised for a new Big Cut. And because it is "only" second growth that is at risk, it is far harder to work up a passionate defense for it than it is for the pitiful, doddering elders of the Hartwick Pines.

I would not want to see all second-home construction, or all logging, stopped. We don't need to turn Michigan into a giant preserve. But we ought to recognize that this reconstructed forest is, by now, nearly as complex and valuable and—yes—natural as the one it replaced. And we ought to demand a great deal more care from the loggers and the developers this time around.

THE AU SABLE RIVER BEGINS AT THE INtersection of Kolke and Bradford Creeks a few miles west of Hartwick Pines, near the tiny town of Frederick. In Grayling, ten miles south of Frederick, the Au Sable is joined by its East Branch, which heads out near Jones Lake. Below the confluence the river is broad enough to float several canoes simultaneously, and it is here, in Grayling, that the Au Sable Canoe Marathon begins. Nineteen hours and 120 river miles later it ends at Oscoda, hard by the Sawyer Canoe factory, where the river is swallowed up by the wide blue sea of Lake Huron. The marathon has taken place each year since 1947, making it the oldest continuously run long-distance canoe race in the country.

The Au Sable is a lovely river, blue and winding between banks of reeds and willows, but it is in no sense wild. Roads run along the tops of the sandy bluffs that line the river for most of its length; bridges cross it in numerous spots. Six times it is plugged by dams. Marathoners portage these, after paddling across the reservoirs. They must also deal with bug hatches, deadheads, masses of spectators, and darkness. (The race begins at 9 P.M., and the first eight hours are run at night.) Occasionally, snow has fallen. More often the temperatures on Marathon weekend—usually the last weekend in July—climb above ninety degrees. Where the river braids into multiple channels, route finding adds to the fun. Some of these extra channels serve as shortcuts. Others, like the notorious Spider Cut fifteen minutes below Grayling, are long, winding waterways that cost precious time. And time is of the essence here. The team that won the inaugural race in 1947, Allen Carr and Delbert Case, took twenty-one hours and three minutes to negotiate the 120 miles from Grayling to the coast. Today's top finishers will complete the same course in under fifteen hours, usually no more than a minute ahead of the second- and third-place finishers. (At Alcona Pond in 1991, more than halfway through the race, the time differential between the leading boat and the fifth boat behind it was less than twenty-five seconds.) The course record, set in 1994 by a couple of Canadian paddlers named Serge Corbin and Solomon Carriere, is thirteen hours, fifty-eight minutes, and eight seconds.

Why the dramatic dropoff in times? In 1997, journalist Margery Guest put that question to Hugh Bissonette, who won the second annual Au Sable Marathon with his brother Bud back in 1948. "Well," drawled Bissonette, "you take a Model A right now and I'll take a new Ford and we'll have a race and see who wins." The Bissonettes had raced in a bright red wood-and-canvas Old Town canoe weighing more than one hundred pounds. Today's Kevlar craft weigh between twenty and thirty pounds, create far less friction in the water, and take bumps and bangs from rocks and logs with less damage. Racing paddles of the current generation weigh eight ounces each. It is technology, as much as training, that has produced racers capable of paddling more than once per second for fourteen hours running.

We were a week too late to witness the Marathon, which was just as well, because what I really wanted to view was the river. We crossed it at Grayling shortly after leaving Hartwick Pines and didn't see it again until Mio, thirty-two miles east. (The name of the town of Mio looks as though it should be pronounced "MEEoh," but locals give it a long I: "MYoh." The name of the river itself is pronounced "oss OBul." I said "AW zabul," the way an identically spelled river is pronounced in New York, and got blank stares. These are what Garrison Keillor has referred to, in another context, as "test names.") At Mio the river backed up behind the first of its six dams and the road crossed from the south to the north bank. A large green lawn sprawled just below the bridge, sprouting canoes and canoeists like dandelions. Mio is full of outfitters, and the run from there to McKinley, fifteen miles downstream, is the most popular day trip on the river.

Two or three miles downriver from Mio along Road F32 we pulled off at a dirt turnout labeled "Au Sable River Overlook." Thirty feet below us the river swept by, lined with sedges and willows. Canoes were attacking it. There is really no other word to use here. Boats were coming around the upriver bend every few seconds. Sometimes they were individual craft with one or two paddlers. More often they came in groups of three to five, lashed thwart to thwart. The people in these boats had paddles in their hands, but it would be stretching it to call them paddlers. Most of them seemed to be there to party, as if that stretch of the river was a long, linear fraternity bash. Loud banter, laughter, and blasts of recorded music floated up through the wincing air. Every minute or so, one of the groups of river revelers would begin yelling loudly and banging rhythmically on their boats' aluminum sides. Others would rapidly join in. The din was enormous. I could see why Rod Badger, who won't get into one, always refers to aluminum canoes as "lumiboomers."

"If this goes on all summer . . ." said Melody. She left the ellipsis unfinished, but I knew what she was thinking. Wild birds and mammals tend to be sensitive to noise. Any left in the area—if there *were* any left in the area—would be experiencing large amounts of stress, which would alter their feeding habits, shorten their lives, and drop their reproductive rates. Such effects tend to magnify through an ecosystem, changing the balance of the plants and insects the birds and mammals would normally feed on. Tourism is often sold to rural areas as an environmentally friendly substitute for logging, but in this part of the river corridor logging might actually be the better alternative.

There is definitely one endangered species in the central part of the Michigan Mitten for which logging, properly carried out, is a boon. The lands around Mio and McKinley are the only known breeding grounds of the Kirtland's warbler, a small, shy, yellow-and-brown bird about the size of a house sparrow. Since intensive study of these birds began in the 1970s there have never been more than five hundred of them; the population is down, now, to just over four hundred. Ornithologists are trying to raise that number, but it is a hard battle to win, and it uses what may seem to a non-ornithologist to be some very odd tools.

We came upon one of those tools just east of McKinley. Road F32 had pulled away from the river a short distance and was running out on the plateau, surrounded by pine forest. On the left side of the road the forest suddenly gave way to an eighty-acre clearcut perhaps twenty years old. The clearcut had been replanted in jack pine; the young trees stood close together, their branches interlocking, their growing tips a uniform twelve feet off the ground. This was a Kirtland's warbler nesting area. The birds build their nests on the ground at the bases of young pines. The pines must be close together, of uniform age, and between six and eighteen feet in height. The picky warblers reject anything that doesn't meet these criteria. If they can't find proper nesting habitat, they simply don't nest.

In presettlement times, Kirtland's warblers relied on natural disasters like the 1940 windstorm that flattened half of the Hartwick Pines. It was a touch-and-go process; the birds must always have been a marginal species. Today, with much of the forest gone and most of the rest under as much protection from natural disasters as we can give it, touch-and-go would be little touch and almost all go if we didn't help it along. Hence these jack pine–choked, weedy-looking clearcuts. To human eyes, they are much uglier than the old-growth sanctuary at Hartwick Pines. But if all Michigan looked like Hartwick Pines, there would be no Kirtland's warblers.

For two hours we closely followed the Au Sable, though we rarely saw it. The road bored through woods, staying a discreet distance back from the banks. At intervals there were side roads to overlooks, and we almost always took these. The increasingly broad river flowed serenely past at the base of its bluffs. We were well past lumiboomer land by now, and the few canoes we saw moved as quietly as the river. Loud Pond, the third of the Au Sable's six reservoirs, seemed badly misnamed: it was much quieter than the stretch of river directly below Mio.

Eventually we came to the coast at Oscoda. The river sought the freshwater sea through a crowded marina full of boats moored at barely deck-bumper distance from one another. This is the fate Max Holden is trying to stave off for the Platte. The marina looked little changed from the day I saw it with Rod Badger fifteen years ago; the Sawyer Canoe plant just upstream, where Rod had purchased the canoe we carried for the rest of the trip, had not altered much, either. We checked into an Oscoda motel and drove south, past disquietingly large amounts of new coastal sprawl, toward Tawas Point.

Tawas Point is a sandspit that hooks into Lake Huron where Saginaw Bay cuts back into the Mitten to form the gap between the Thumb and the fingers. Low dunes run behind a long crescent of beach, backed by large freshwater marshlands. The point is a complex, fascinating structure which often appears in geology textbooks as a type example of a hooked spit. Most visitors here, however, don't come for geology. Most come for—

Well, to be perfectly honest, I have never been able to figure out what they *do* come for. I know why *I* come: it is to walk what is, hands down, the loveliest beach on Michigan's east coast. That, however, does not appear to be a terrifically common objective. When Rod and I were here in 1983 we were quite literally the only humans on the beach, though the campground less than a quarter mile away was full to overflowing. This year there were a few more people, but it was still not what you could call crowded. We walked south along the spit in the waning light, watching the surf curl up and purr on the sand, wading up to our knees in the cold, clear water. Up on the storm beach, near the base of the dunes, gulls were arguing themselves down for the evening. A semipalmated sandpiper dashed and statued along the rift-darkened sand at the edge of the infinite sea.

Perhaps it is simply that I have come both times at twilight. Perhaps the beach is Coney Island–full in the heat of a summer afternoon. Whatever it is, it left me appreciative but puzzled. "E.T. phone home," Rod had said last time, as we returned from the empty beach to the urban rush and stress of the overflowing camp. That sense of being marooned on an alien planet full of distinctively oddly behaving creatures still describes beachwalking on Tawas Point on a summer evening. Other beaches are full of sunset walkers; where are they here?

Darkness came down over the waterfowl-filled marshes behind the dunes. We came back to the car—it was almost the last one left on the large lot—and drove north, past the endless parade of brightly lit beach houses full of people trying to get away from it all by bringing it all with them, back to Oscoda and our motel.

WE SPENT ALL THE NEXT MORNING DRIV-
ing around Saginaw Bay, and we didn't stop all that often, either. The
bay is 60 miles long and 20 miles wide at its narrowest point, and the
distance around it, from Oscoda on the northwest side to Grindstone
City on the southwest side, is more than 150 miles.

Saginaw Bay is shallow, like Green Bay, and like its Wisconsin
counterpart it is subject to natural eutrophication. Both bays trend
northeast-southwest, and both have cities of roughly 100,000 people
(Green Bay, Wisconsin; Bay City–Saginaw, Michigan) sitting astride
the mouths of rivers draining into their south ends. Both rivers drain
rich agricultural areas; in both cities, the riverbanks are lined with
industry. At about this point, however, the similarities stop. Green
Bay is long and slender and closed off at both ends; were it not for
the narrow, shallow opening of the Door halfway up the east side,
it would be a separate body of water. Saginaw Bay is shorter and
dumpier, and its mouth is wide open to Lake Huron. Green Bay is
hemmed in by bluffs on the west and by the three-hundred-foot-
high limestone cliffs of the Niagaran Cuesta on the east; Saginaw Bay
is entirely ringed by sand and marsh, and the land is low and flat.
The Door Peninsula between Green Bay and Lake Michigan is filled
to overflowing with tourists and new housing developments. The
Thumb—well, the Thumb has some new development along part of
its shoreline, and it is trying to attract tourists. In some places it may
be trying a little too hard.

The morning was still young when we reached Tawas City, a quiet
town tucked in behind Tawas Point on the north shore of Saginaw
Bay. We stopped briefly in a waterfront park. There were a couple
of dog-walkers out; boats in the nearby marina rocked and creaked,
and little wavelets splashed up the sand. Drifting waterfowl played
peekaboo among the mists rising from the morning water. The town
felt maritime. Most towns on the Great Lakes do; perhaps that is

because most towns on the Great Lakes are. The Lakes have spawned a seaboard culture that has far more in common with Maine or Oregon than it has with the breadbasket states that are commonly joined to it in the geography books. Great Lakes residents may not recognize their regional bonds, as Jane Elder had worried back in Wisconsin, but the bonds are nonetheless real. They, and we, must stop thinking of this area as part of the Midwest: it has little more in common with Iowa or Missouri than it does with the back side of the Moon.

Between Tawas City and Bay City, down the north and east sides of the bay, the shore was lined with displaced city suburbs in the pattern that had become all too familiar elsewhere in the Great Lakes Basin. Then we turned the bottom of the bay and headed out the Thumb on Highway 25, and the suburbs suddenly stopped. Fields of sugar beets spread out over the flat floor of ancient Lake Saginaw. Roadside ditches were bright with the purplish-red blossoms of loosestrife. In some places the loosestrife was fronted by rows of bright chicory above a thick floor of morning glory, a red-white-and-blue effect that, though happenstance, looked as though it had been planned by a landscaper with a strong patriotic streak. Marshes lined the bay, musical with birds.

At intervals the highway passed through tiny little dying towns. Most seemed pretty well shut up this Sunday morning. Bay Port, halfway up the inside edge of the Thumb, had banners above its streets advertising a Fish Sandwich Festival: booths lined a small city park, but there were no customers at them. The streets were eerily empty. Perhaps it was the hour; it was still a little before noon. I couldn't help, though, contrasting the scene with the milling madhouse of visitors that had seized the streets of Fish Creek, back on the Door. Where were the crowds of fish sandwich lovers the festival was aimed at? Wherever they were, they weren't stopping in Bay Port. I think I shall always regret that we didn't stop, either.

At Sand Point, near the mouth of the bay, there was a sudden change in scene. Sand Point is a long, skinny peninsula stretching almost due west into the bay's shallow water. On the south it encloses Wild Fowl Bay, a great open marshland whose big sky is almost constantly filled with waterfowl. On the north it forms the west end of one of the longest continuous beaches on Lake Huron, stretching more than twenty miles around the tip of the Thumb to Port Austin, and though the marshes and beet fields below the point had been nearly devoid of homesite development, here it was thick. The low dunes behind the beach grew shoulder-to-shoulder houses. Every few dozen yards, great metal frames sprouted out of the surf; these frames held powerful motorboats suspended from cables. People swarmed the sand. None of them seemed to have fish sandwiches in their hands.

Beyond Port Austin, as suddenly as it began, the strip of sprawl ended. The highway stayed a chaste distance back from the edge of the Lake, which was lined with low cliffs of Marshall sandstone, six to ten feet high and undercut by wave action. Marshall sandstone is a close-grained stone of surpassing toughness; early Saginaw loggers made grindstones out of it. Many of these grindstones, three to six feet in diameter and up to a foot thick, still lie about the northeastern portion of the Thumb. They often carry "For Sale" signs. A nice item for your yard back home, but the weight would probably destroy the suspension of a half-ton pickup; with the Escort it wasn't even a close call. We came into Harbor Beach, still grindstoneless, and grabbed lunch at a locally owned hamburger stand.

Between Harbor Beach and the base of the Thumb, Lake Huron is lined by sand bluffs with narrow beaches at their feet. Here the houses started again. We dashed past, stopping only once, briefly, at a small blufftop park with a view over the wide water. I was getting anxious about the evening, which, due to Michigan's merciless two-night rule, we were once more facing without reservations. The logical camping spot would be Lakeport State Park, a few miles from Lake Huron's outlet into the St. Clair River; after our recent experience at Hartwick Pines, though, I was not at all certain there would be any sites left when we got there. Half an hour above Lakeport, near Port Sanilac, we made a quick run through a county park campground. It was not encouraging. The "campsites" consisted of open lawn on which campers and RVs were parked cheek by jowl, close enough that a few tin-can telephones could have served the entire crowd as a communication network. We Forded back onto the highway and flew toward Lakeport, hoping for a miracle.

We got it. The park, with its wide green lawns and rustic old park store—among the last such stores in Michigan—was barely half full. We picked out a site on a little-occupied loop, put up the tent, and spent a long time watching a pair of Canada warblers feeding a clutch of young in a nest on the ground in a small woody area between the next two loops over.

After a while we wandered down to the Lake. Clouds had begun to gather back at Harbor Beach; by now the sky was covered with a light overcast, muting the evening inching slowly upward from the eastern horizon. The beach was sand, accented by a light scattering of pebbles. Here and there among the pebbles were thumb-sized, lightly striped white shells. Zebra mussels. After ten days of searching along more than a thousand miles of freshwater coast, I held the scourge of the Lakes in my hands at last.

I have to confess I was not very impressed. Later, in Lake Erie, I would be brought face-to-face with evidence of the havoc these tiny molluscoid zebras have been wreaking for the last ten years; here, less than fifty miles north of the point where the first of them had been pulled from Great Lakes waters—Lake St. Clair, back in 1988—they seemed strikingly innocuous. There were only a few of them, and they were very small. Rather attractive, actually, among the damp pebbles at the edge of the water. Knowing what they were and what they meant, I photographed a few, and collected a few more; but I could work up no sense of urgency about them. The clouds cleared away from the Moon, high in the southern sky. The distant lights of the International Bridge across the St. Clair River, at the mouth of Lake Huron, danced on darkening water under a still-bright sky and one lone, unblinking star.

On our way back to the tent we passed a small group of children and young teens, all black, standing at the edge of the Lake. A short distance a way, watching in a relaxed but intent manner, stood a couple slightly older than us. Grandparents and grandchildren, I thought, and then immediately revised that: the kids didn't look related to each other, and the way they were bunched tightly together suggested uncertainty in the face of the totally unfamiliar. Though they were laughing loudly as the waves splashed about their knees, the laughter carried an edge of hysteria. The only breakaways were a pair of young boys who made quick dashes to pick up pebbles; they returned to the safety of the group before launching their stones, *plop!*, as far as they could throw toward the edge of the world. A group of inner-city children up from Detroit on an outing? A church project? A school? I never found out.

I did speak briefly to the couple chaperoning the children. "Isn't it odd," I said, "how when you bring kids to water, the first thing they do is throw rocks in it?"

The man smiled: the woman laughed outright. "Isn't that the truth?" she said. But then they turned their attention back to the children, away from the white guy—me—who felt he had no right to press the conversation further. We returned to our tent. The warblers were still feeding their young, preparing them for life in a world that often seems to care far more about warblers than it does about inner-city children—a deeply troubling situation for those of us who happen to care very much about both.

The group we overheard the next morning at breakfast probably didn't care too much about either one. The restaurant we had chosen met all the right criteria—gravel parking lot, small sign, lots of calendars—but it seemed perversely bent on making sure we understood that those were guidelines, not guarantees. The coffee was passable, but the food was among the worst we encountered on the entire trip, and who am I to say that the state of cleanliness in the restrooms was responsible for the stomach cramps and diarrhea we both suffered later that day? I will not name the place, not from charity but because we quickly and mercifully blotted its name from memory. But I cannot erase the memory of the food, or the snatches of conversation drifting over from two tables away.

" . . . give me the bill and I'll introduce it . . ."

" . . . the good-old-boy network was hidden. We're taking comments . . ."

" . . . without violating the zoning ordinance . . ."

" . . . I'm the supervisor. I can just do it . . ."

They were all white men, all in their fifties and sixties. I can only guess that one was the local state representative and that the rest were developers. It is not a guess to say that there was a great deal of money at stake, because they talked about that. More shoreline was going to be lost; a bill in the legislature was going to chip another chink out of the legal protection for the rest of the coast. Meanwhile, less than five miles away, Lakeport State Park was deteriorating for lack of funds, and a pair of warblers and a pair of concerned humans were each trying to get a group of at-risk children safely launched into a cold and largely uncaring world. The Bible says judge not that ye be not judged, so I will not judge the men at that restaurant table. But their priorities were certainly far different from mine.

We paid for the poor breakfast and walked out into the hazy early-morning sunshine. We were in the land below The Line, now, and headed deeper into it. Another nine days would pass before we would be able to escape.

V
BELOW THE LINE

Take any map that shows all five Great Lakes—the one in the front of a road atlas will do—and lay a ruler across it so that its edge passes through Green Bay, Wisconsin, in the western part of the map, and Kingston, Ontario, in the eastern part. Run a pencil along this edge. You have just created a reasonably accurate representation of The Line, a demographic boundary which divides the Great Lakes Basin into two separate and distinctly unequal halves. Above The Line is the outback—timber, rock outcrops, rushing streams, parklands, and only a scattering of people. Below The Line is a great clot of industry, the largest concentration of people and production facilities in all of North America.

The existence of The Line was noted as early as 1959 in a paper by two geographers, Andrew Clarke of the University of Wisconsin and Roy Officer of Waterloo University in Ontario. It remains every bit as impressively distinct today. A striking representation of it may be found in the U.S. Environmental Protection Agency's *The Great Lakes: An Environmental Atlas and Resource Book,* though it is not labeled that way. The atlas contains a map of the Great Lakes Basin titled "Employment & Industrial Structure." On this map, all metropolitan areas of more than 100,000 people are represented by red circles, with the diameters of the circles drawn proportionate to the total population of each area. Above The Line there are three small, widely spaced circles, representing Duluth, Minnesota; Thunder Bay, Ontario; and Sudbury, Ontario. Below The Line there are twenty-four circles, and they are so large and so close together that the whole southern section of the map is essentially red. A line drawn tangent to the two northernmost circles in this blood-red splatter—Green Bay and Toronto—will diverge from The Line by only a few degrees.

The Line owes its existence to two—or possibly three—closely related physical phenomena. Above The Line, the land is a huge jumble of glacial landforms: moraines and outwash deposits to begin with, polished knobs and ridges of bedrock further north. Below The Line there are some of these—the southern edges of Lake Erie and Lake Michigan are defined by them—but most of the countryside is covered with alluvial soils, dropped there by ancestors of today's Lakes—Saginaw, Whittlesey, Chicago, Algonquin, Iroquois—as they succeeded one another along the front of the steadily retreating ice. Related to these differences in soil are differences in plant growth: the boundary between the mixed-hardwood forest of the middle latitudes and the coniferous forest of the North, known to ecologists as the "oak-pine ecotone," is almost exactly contiguous with The Line. These congruities were noted by Clarke and Officer. They did not mention a third, equally striking lineament: the hinge line. This is the boundary drawn by geologists between the stable south and the isostatically rebounding, slowly rising north. It may be purely coincidental, but this boundary, too, follows The Line with near-mathematical precision.

If you have drawn The Line following my instructions, you have probably noted that we actually had stayed below it for most of the trip so far, sticking our noses out only on the Door Peninsula and in the vicinity of Traverse City. Except for the run around Lake Michigan's South Basin, though, we had been sticking pretty much to its northern fringes. Now we were heading into the heart of it, around Lake Erie and Lake Ontario. Our route would run precisely opposite my course in 1983. Then, I had followed the southern shores of the two smallest Great Lakes westward through Rochester and Buffalo and Cleveland and Detroit to the St. Clair River and Lake Huron. This time we would leave Lake Huron via the St. Clair, then follow the Erie and Ontario shores eastward.

It was nearly 10 A.M. when we crossed the International Bridge into Canada. We cleared customs quickly and headed south from Sarnia along the St. Clair, through the area known to Canadians as Chemical Valley. Immense petrochemical plants lined the road on the east side; little parks dropped toward the water on the west, their green lawns ending above concrete seawalls. Six-hundred-foot lakers dieseled majestically by, their pilots engaged in a task roughly equivalent to threading two football fields joined end to end through a twisting course not much more than two football fields wide. Upriver and downriver boats passed each other with only yards to spare.

In Wallaceburg we got a small taste—quite literally—of life below The Line. We had stopped at a very local, decidedly non-chain fish-and-chips stand for lunch. The angular, thickly breaded filets were delicious. Halfway through eating, though, my eye fell on a copy of the London, Ontario, newspaper that someone had left lying about. The paper contained an article identifying Wallaceburg as the last city still obtaining its drinking water from the St. Clair River. More than fifteen hundred man-made chemicals have been identified in the St. Clair's water. About ten years ago there was a movement afoot to connect Wallaceburg's water system to either Lake Erie or Lake Huron. The movement fizzled: the group pushing it disbanded. The chemical-laden water continues to flow. Chemical pollutants being the long-term time bombs that they are, I would still place my money on breakfast—and biological contamination—as the cause of the upset intestinal tracts we would suffer a few hours later. But I'm not sure I want to look at my liver and kidneys in twenty years.

South of Chatham, below Canada 401, we broke out once more into rural terrain—the flat, fecund region Canadians refer to as the Sun Parlour. The highway angled southeast over the bed of ancient Lake Whittlesey; fields foundered under the rich weight of harvest-ready tomatoes. In twenty miles we came to Lake Erie, but we didn't actually reach it. We could see it—a horizonless field of blue behind scattered blufftop houses—and we could hear it smashing against the bases of the bluffs each time we stopped. But we couldn't get to it. As bad as the United States is at providing public access to the lakeshore, Canada is worse. The parklands along the St. Clair River are an anomaly. In most of Canada's part of the Great Lakes Basin, if you want to go to the beach you have to buy it first.

This general lack of public access probably contributes to the crowding at Point Pelee—but Pelee is very small, and very special, and it would probably be crowded anyway. At barely eight square miles, Point Pelee National Park is the smallest in Canada. It is also among the most heavily used, with between 400,000 and 500,000 visitors annually. Those visitation numbers, however, are actually down significantly from the park's peak years in the early 1960s, when nearly 800,000 people crammed into it each year, mostly on summer weekends. The reasons bear some scrutiny.

The Peeled Point—that is what *Pelee* means, in French—is a long, triangular sandspit extending some fifteen miles into Lake Erie south of Leamington, Ontario. Geologists disagree about its origin. At one time it was thought to be the north end of an underwater ridge extending across the Lake to Lorain, Ohio, but recent studies indicate that this ridge actually begins at Pelee Island, eight miles off the point to the southwest. The truncated mound that forms the substrate of the point is definitely glacial in origin, but it is unclear how it was deposited, although the shifting drainage patterns as the ice melted back toward Hudson's Bay probably had something to do with it. The point itself—the part visitors see—is sand, sculpted by the currents of Lake Erie into a complex dune-and-spit structure with two principal berms. Where the berms come together to the south, a long, curved spit called the Tip extends well into the Lake. Where they open away from each other to the north, an immense wetland, the Great Marsh, lies between them.

The east berm is low, and much of it is usually covered by water. The west berm is higher and better stabilized, and on it grows a forest like none other in all of Canada. There are citrus trees, honey locusts, sycamores, hackberries, sassafras. Grape and Virginia creeper climb among them. The place looks as though it had been plucked, intact, from the coast of the Carolinas—an impression which is not as far-fetched as it seems. This is, in fact, a Carolinian forest. Its species mix is like that of the Eastern Woodland biome of the United States, a biome once known as the Carolinian, and its type occurrence is in North Carolina.

Pelee's special mix of broad marsh and anomalous woodland, combined with its strategic position on the bird-migration path known as the Eastern Flyway, keep the park pretty much filled with birds year-round. More than 360 species—including 41 of the 55 known species of North American warblers—have been sighted within this tiny area. Birding is good year-round, but it is especially spectacular during the peaks of the spring and fall migrations, as birds alight on the point to rest before and after their long, exhausting flight across Erie. As many as twenty-five thousand birders visit Point Pelee in each of those seasons, and their presence pumps several million dollars annually into the local economy.

We came into Leamington from the east and turned south toward the park, beneath banners announcing an upcoming Tomato Festival. Houses gave way to fields, then woods and wetlands. A light haze softened distances but did little to temper the heat or brightness of the midafternoon sun.

At the Marsh Boardwalk, a mile-long circle of wooden walkways, we strolled dry-shod into the heart of the Great Marsh. Carp nosed about in the shallow water; turtles sunned themselves on clumps of grass. Red-winged blackbirds perched on reeds, converting them-selves to song. Thousands of acres of cattails sprawled toward the horizon, accented here and there by small clumps of trees. This is

Canada's Everglades, a vast wetland occupying the nation's south-ernmost extension, and the look and feel of the two places is strikingly similar. The man I overheard asking his small daughter if she had seen any alligators yet was probably joking, but Pelee is the kind of place that ought to have them.

Three miles past the Marsh Boardwalk the park highway ends at a small, crowded information center. Here visitors transfer to the Balade, a rubber-tired train pulled over a narrow paved road by a propane-powered pickup. The Balade, which has been in use since 1972, is the centerpiece of Point Pelee's visitor-control policy. Back in the early sixties, parking lots covered much of the Tip, and they were crowded to overflowing with cars. Park officials determined that most of the cars were there, not because of Pelee's special pleasures, but because there were parking lots next to a beach. By removing the lots and closing the road to cars, they filtered those who simply wanted easy access to sand and surf out of the visitor stream. Those

The Great Marsh at Point Pelee, Ontario

who remain are here because they want to see Point Pelee. A clause in the park's operating plan allows the gates to be closed when visitor densities get too high, but that clause has rarely been invoked. Usage has climbed, but only very slowly: the crowds seemed little different in 1998 than they had when Rod and I came through, on a similar day at the same time of year, in 1983.

The sandy Tip curved south like a scimitar into Lake Erie. Waves crashed against the east side; little wavelets lapped softly on the west. Pelee Island was a smudge on the southwest horizon. Small specks moved in front of the island, in Pelee Passage; binoculars resolved them into six-hundred-foot freighters. People stood at the edge of the surf and had their pictures taken on the southernmost piece of land in mainland Canada. A few hundred yards up the point, back near the place where the Balade had dropped us, a large sign identified the location of the forty-second parallel: much further west, that parallel becomes the northern boundary of California. We who dwell in the continental United States are used to thinking of Canada as the Far North, and it is not easy for us to recognize that part of it lies as far south as the Golden State, south of our own house in Oregon. But here we were. The pulse of people that had arrived with us on the Balade dwindled and thinned; for a few moments before the next pulse arrived we were alone on a beach at the end of Canada, on the shore of the limitless sea.

On our way out of the park we stopped at a marsh overlook beside the highway and pulled out the binoculars for one final look around. Another couple joined us, also with binoculars. The man spoke with a heavy German accent; the woman spoke only in German, and not to us at all. But the conversation was a familiar one to birders everywhere:

"Are you seeing anything?"

"Not much, this time."

"Is that a heron?"

"I didn't see one—wait! Yes it is! That's a great blue."

We exchanged satisfied smiles.

A half hour later we were in Windsor. Traffic franticed past; across the river to the north (yes, the *north*) squatted the square, muscular towers of downtown Detroit. We crawled into our motel for the night, shutting the door against the city, holding the memory of Pelee like a magic stone in our tightly clenched hands.

THE HEADQUARTERS OF THE INTERNA-
tional Joint Commission on Boundary Waters is in Windsor, hard by
the southern portal of the Detroit-Windsor Tunnel. Across the wide
Detroit River looms the Renaissance Center, the soaring seventies
symbol of the Motor City's rebirth. The IJC's mandate covers all
waters shared by the United States and Canada—the Skagit, the
Columbia, the Milk, the Red, and others, large and small, in every
state and province along the border—but its principal focus is the
Great Lakes. More specifically, it is the Great Lakes Water Quality
Agreement—the GLWQA.

The GLWQA is a child of the sixties, though it wasn't signed until
1972 and it actually grew out of the Boundary Waters Treaty of 1909.
It was cobbled together in the wake of the widely reported "death"
of Lake Erie, and in its original form it simply committed the United
States and Canada ("the Parties") to clean up the Lakes. Little was
said about how this cleanup was to occur, or by what standards the
Lakes would be judged to have become clean.

In 1978 the GLWQA was heavily revised, and the almost reveren-
tial status accorded to the Agreement by Lakes activists stems from
the changes made that year. The 1978 Annex to the Agreement was
built around two radical ideas, each of which can be expressed in
two words: *ecosystem approach* and *zero discharge*. The first meant
that the Parties had pledged themselves to seek solutions which
would treat the entire Great Lakes Basin as a single system, without
regard to political boundaries. The second referred to persistent toxic
substances—materials such as DDT and PCBs—and it meant that implementation of the Agreement would not be considered successful
until there were no longer any releases of these particularly insidious
pollutants into Great Lakes waters at all.

In 1987 the Agreement was revised again. Environmentalists
swarmed the hearings on these revisions, fearful that the ecosystem

approach and zero discharge would be flushed down the St. Lawrence River. That didn't happen, but something else did. Back in Madison, Jane Elder had complained that in recent years the once-robust IJC seemed to have turned wimpy. Here in Windsor I was about to find out why.

On my first trip through the Lakes, midway in time between the 1978 and 1987 Annexes, I had spoken with an IJC staff person named Pat Bonner. Pat was no longer with the agency, but the lack wasn't critical: the man I was going to see on the morning after Pelee was someone I had been acquainted with nearly as long. I had first met him in 1987, at the same Great Lakes United conference in Niagara Falls where Jane and I had taken the now-famous tour of Love Canal. His title with the IJC was simply "environmental scientist," a phrase with a wide degree of latitude to it. His name was John Hartig.

John occupies a corner office on the eighth floor of the IJC's building on Oulette Avenue: his desk is positioned to face the expansive view north and east over the Detroit River. That meant, of course, that when we sat down across the desk from him our backs were to the view. Melody warned him we might be turning away from him a lot. John laughed. He is tall and sandy-haired, with a youthful look that belies his twenty years with the commission as thoroughly as his title belies an eighth-floor corner office.

"I think the big change in the last fifteen years," he told us, "is that there's been almost a revolution of government. A *de*volution of government. You have federal governments transferring roles and responsibilities down to state and provincial governments, who in turn are saying, 'We can't do it all, so let's throw some responsibility down to local and regional governments.' The big concern that I would have would be, how do you ensure that these lower levels of government—if you can call them lower levels of government—don't collapse under the weight of all this new responsibility? I think it has enormous ramifications. Some things are not getting done."

"Well," I said, "one of the big things is that the RAPs are underfunded."

"And RAPs—they're just *one* example. That's *one*."

"So do you want to talk about that?"

"Sure." He composed himself. "You have some governments—the state of Michigan is one of them—who say, 'Our primary responsible is the issuance of permits for water and air. We are mandated to do that, through the Clean Water Act and the Clean Air Act. These peripheral programs are nice, and would be great to have, but we just can't place a lot of priority on them.' They come at it from very much a command-and-control, top-down philosophy. And I would say that that's had enormous ramifications. Michigan has gotten rid of all but one of its RAP coordinators. Should government have a role in some of these bottom-up, community-based approaches? In my opinion, they *should* have a role. Government should participate in the decision-making process. If it's a permitting issue, they can get involved in that, but they should come at it from a broader perspective, not just permitting. If you do that, you miss things like habitat, and biodiversity, and exotic species—whole things like that.

"Look at Humbug Marsh." Humbug, a few miles below Detroit, is the only large wetland remaining on the American side of the Detroit River. "The Michigan DEQ [Department of Environmental Quality] should be able to see how important that is. They have a public trust responsibility to protect it. But they're so preoccupied with the permits, and doing all this stuff, that you've got the *jewel* of western Lake Erie and the Detroit River— and they're gonna put a $500 million development in it. So you could miss whole things. How do you make sure that there's a balance? That you look at all the issues? *Who* looks at all the issues? Governments today are saying, 'Well—it's not our job. We'll leave that to somebody else.' And in most cases, the somebody else isn't even *defined* yet.

"So if I could say something," he went on, a little more quietly, "it would be, maybe we should do some work on clarifying roles and responsibilities. Ontario's done the same thing— they've gotten rid of much of their RAP work. They've lost, in some cases, up to 60 percent of their human and financial resources for some of these programs. That's, that's—that's *gutting* a program. They call it 'efficiencies.' But it's actually a real gutting of programs."

" 'Leaning down,' I think, is one of their philosophies," said Melody. She laughed. "But that's not it."

" 'Leaning down,' John quoted, and he laughed, too. "I've got to be a little careful with this stuff."

"I understand," I told him.

"But it's really a pendulum. They thought the pendulum was way off to the left, so you had governments that came in and said, 'The pendulum needs to swing.' And the pendulum has swung. It's swung way beyond the middle, and it's way off on the right. Now it's got to come back. Look at Great Lakes monitoring and surveillance. You know, we were a leader, globally. Everyone looked up to us, to the way we did phosphorus. But now they're saying, 'Phosphorus is over, we don't have to worry about phosphorus, so we don't have to do that anymore.' So they're not doing it. And now they're trying to do these LaMPs without an adequate database, without the knowledge, without the understanding, maybe, or the current information that they need."

"There's still some monitoring work going on," I pointed out. "I was just talking to Bob Beltran of the EPA, and he was getting ready to go out on their research vessel."

"Yeah," said John, "in certain areas we're still a leader. I would argue that the Green Bay Mass Balance Study and the Lake Michigan Mass Balance Study have global significance. Out of that has come a greater appreciation for the fact that we haven't finished the job yet. They've shown that we need to set relative ecosystem priorities. So in that case, we're a leader. But—we used to do that on *all* the Lakes. *All* the tributaries. *All* the connecting channels. We had reports that were produced on the state of the Great Lakes. Every two years my life would—get turned upside down"—he shook his head—"to go through that, at that level of rigor. We had a committee for each of the Lakes. We had a Lower Connecting Channels Task Force; we had an Upper Connecting Channels Task Force. They would have two years, and they'd get the charge—'Tell us about the state of the Lakes. What are the key problems? What are the next priorities from a management perspective, from a research perspective, from a monitoring perspective?' And that has sort of gone away."

"That's money, isn't it?" I asked.

He nodded. "That's money. That's money and time—and people. People are money, I suppose."

"In Madison," said Melody, "we talked to someone from the Wisconsin DNR who's doing biodiversity. And he talked about *beautiful* reports. And we asked about Lake Michigan, and he said, 'Oh. Well, our priority is the *inland* lakes.' That feeds right into what you're saying about priorities."

"Yeah," John agreed. "Just look at Lake Erie—how much it's changed with exotic species. You know, on a scientific level, we could have a great discussion on how much we *don't* know about Lake Erie. We say, 'Well, here are some options that we have to better manage Lake Erie.' But the modeling to predict what effect each option would have—we don't have that."

"That's what Bob's out there trying to learn," I said. "But, you know—it's one vessel."

"Yeah," John agreed. "It's one vessel. Again, it's the level. Used to be, you'd look at Lake Erie, and you had Ohio State—they had their whole program. You had CLEAR, the Center for Lake Erie Area Research, and they had a couple of vessels. You had the Ontario Ministry of the Environment. So it wasn't just one vessel and one agency. It gets back to the cuts in state and provincial funding, as well as federal funding. Because it's the combined effect, where you've lost the capacity to monitor and do surveillance."

"The take on the IJC," I said, "and I'm going to throw this at you without telling you whether I agree or not—"

"OK."

"—is that it has become heavily politicized. That it used to be a leader as an agency for scientific management, which is rare among resource agencies. And now it's all political appointees, and the work is not getting done because it's being sidetracked by the political process. That's a take I've heard from at least two people."

"OK," John repeated. He placed his palms together professorially. "The commissioners have always been political appointees. There've always been three appointed by the prime minister and three by the president. But I think that the 1987 Protocol to the Great Lakes Water Quality Agreement changed things a lot. And you have to be careful how you say this, now—"

"I understand."

"—but not for the better. OK? You used the term, 'the IJC was a leader.' A leader in a lot of areas. I mean, a *champion* for Great Lakes issues. We were out in front on a lot of these things. Well—that was almost all pre-'87. In the '87 Protocol they said, 'We're gonna change the roles and responsibilities.' The primary role and responsibility of the IJC—post-'87—would be review and evaluation. Review and evaluation. They got rid of all the infrastructure that the IJC had put together—the Lake Committees, the Connecting Channels Task Forces, all this work—it's all gone. All that stuff is gone. The responsibility went over to the Parties. The IJC is now supposed to do review and evaluation. I think that had a big effect on how we work together, binationally, and how we do this sort of big-picture, ecosystem-based management type of thing. I think we lost a lot. And it has—be careful how you say this, but I don't think it's been picked up elsewhere."

"It's not only government agencies that have been losing money," I pointed out. "We were talking with Jane Elder over in Madison, maybe a week or ten days ago. She's no longer with the Sierra Club, as you're probably aware."

"Yeah."

"I asked her why, and she said, 'You really want to know? Down-sizing.'"

"OK," John said.

"I don't know how Great Lakes United is doing."

"GLU is much smaller, and not nearly the force it once was."

"So I think the organizations are pulling back as well."

"Absolutely."

"So if you don't get it from the private sector *or* the government sector, where are you getting it from?"

"Let's talk about that," said John. "I'll bring up Humbug again. Let's say we were out there. And you say, 'Well, where is the National Wildlife Federation on this? Where is Ducks Unlimited? How do we get them involved?' Well, they have to set priorities—they can't do everything, and they can't be all things to all people. The National Wildlife Federation, in the Great Lakes, has a primary interest in toxics, and not necessarily in biodiversity or wetlands preservation. That's a simplification, but it's true. They've downsized a little and they don't have, maybe, what they did before in some of those groups, so their capacity to do things has shrunk. We have *pockets* where things are going well. But it's really dropped off. It's a real art form to make it all work. That would be my twist on that. But the Protocol has had a huge effect."

"The other thing that happened in the Protocol," I said, "is that the RAPs were created."

"Created and officially sanctioned. Yeah."

"So they create more work, and then they take the—"

"Right," said John. "And in some places, you know, some of these governments resent it. They call it an 'unfunded mandate.'"

"Right," I agreed.

"Right. And they can't stand it. You can pick out the governors and premieres that think like that. But you also have places like Ohio, where *all* the RAPs are going well. Every RAP in Ohio is going well. There are some really good models out there of how to do business in this new era. Hamilton Harbour's going great. They've put $21 million into habitat rehabilitation and contaminated sediment remediation. It's a leader."

"Fifteen years ago," I observed, "Hamilton Harbour was considered one of the two or three worst spots on the Lakes."

"Yeah," John agreed. "So—you know, I'm a half-full-glass person, so to speak. Instead of saying we have all these problems, I'd point to

Collingwood, and Hamilton, and Rochester, and Cleveland, and the Rouge—"

"Well," I interrupted, "Collingwood's the only one that's been delisted."

"It's the only one," he said, not sounding half full at all. "The *only* one."

"Is there a reason for that? Why were they successful when others haven't been?"

"Well, first of all, they were a test case early on. One of the first RAPs to be reviewed was Collingwood. It came to the Water Quality Board, and they rejected it soundly. They said, 'This is the same old way of doing business. You have got to change.' That was back in '85 and '86, so that was—"

"That was pre—"

"Pre-Protocol. Right. So they went back and changed their whole RAP process at that point, and made it community-based."

"Officially," I said, "we didn't even *have* RAPs at that point. Isn't that correct? The AOCs were in place."

"Well—we had RAPs. The Water Quality Board recommended in '85 that they do this comprehensive and systematic RAP process. They made a verbal commitment at that point. They didn't know how to do it—nobody really knew how to do it at that point, to be honest—but then in '87 it was mandated in the Protocol, and it became codified. But the first four out of the chute all started immediately in '85. The Rouge River in Michigan—the Fox River and Green Bay—Hamilton Harbor—and Collingwood. So they were two years into it by the time the Protocol came along. The Protocol was really good, because it codified it and made it official, but the train was already moving at that point."

"Well, let's talk briefly about the Fox River," I said. "Because, of the four you mentioned, that's the one that's still having serious problems, to the point where they're talking about bringing in Superfund."

John sighed heavily. "Yeah," he said, "some of these—if you go around and look at the significant contaminated sediment problems—the *price* tags. Kalamazoo is going to be upwards of $500 million. The Fox River is gonna be $500 million to a billion. *Huge* numbers. So who should pay for this? Five and ten years ago, we had all these polluter-pays laws. If you did something, and we could prove it, then you had responsibility for taking care of it. And in some cases those laws have been gotten rid of, and in some cases they're not being enforced to the fullest extent. The other angle on that is, you could go into litigation. Waukegan is one of our major success stories. They've had an order of magnitude decline in PCB contamination in fish. They've eliminated the health advisory for fish in the harbor. The marina is now 100 percent filled—it was only 40 percent filled before. Property values have gone up. Finally, there's some payback. But the message in Waukegan was—this is why the Fox is so afraid—fourteen years of no action. Fourteen years of litigation. Between attorneys and consultants, they spent about $14 million. In litigation, the only ones that win are the lawyers. The environment does not win, and I don't think the *people* win in that kind of situation.

"So it was a big lesson. Does everyone want to go into this protracted litigation? In the end, you haven't necessarily decreased the amount of remediation—you still have to do the same level. But, to your surprise, the costs have gone up substantially. So after spending $14 million in litigation, they had to put out a check for $23 million for sediment remediation. I think, obviously, the Fox River is afraid of the magnitude of the dollars. And now, maybe the threat of this Superfund action might move it along. Each one's different, you know. In Ashtabula all the industries have formed their own partnership and created a pot of money. Put their dollars in place to minimize the litigation, and minimize some of the other costs that occur. They're saying, 'We know we're gonna have to pay something, but we want to make sure it goes to the right thing.'"

"Cleanup instead of lawyers," I said.

"Exactly," John agreed. "Cleanup instead of lawyers. And they get the public relations they deserve for taking care of the problem they created. But the Fox River—that is a huge issue, because of the magnitude of the problem. You know, that's the highest density of pulp and paper mills in the world. And that problem there is so overwhelming—you've got other issues in the Fox River and Green Bay, like habitat, and beach closings, and there's still some nutrient enrichment problems, but the PCB problem *dwarfs* them. And the complexity dwarfs them, with the dollars it's gonna take to solve it."

"So do you think contaminated sediments are more of a problem than something like, say, urban sprawl?"

"I wouldn't say that." He smiled. "I think urban sprawl's one issue we haven't even scratched the surface on. Every time we put in another connector we create another avenue for them to go out further. And when they put in these developments, like the one at Humbug, they don't want to go to a brownfield. They want to go to a greenfield. If you're gonna invest a half a million to a million dollars in a home, you want the best. I think transportation planning is driving that. Some national changes have to occur. You look at the growth of Toronto into that big agricultural belt, and you just wonder if a hundred years from now they're going to be taking subdivisions out of there to put farms back in. Because what's below those homes, and all that concrete, is phenomenal agricultural land. It's the best in Ontario. And yet—Toronto keeps growing. I think that's a real challenge. Yeah, the growth issue's huge. Did you see the casino?"

I blinked. "Not really."

"Oh, my gosh," he said. "It's just a block and a half from here."

"It's all the construction, I guess," said Melody.

"We must have driven past it." I tried to remember.

"Three hundred and fifty million in government funding." John looked grim.

"Government funding," I said.

"Government is—in the business," he said. "Of casinos. The government owns and operates—well, they subcontract some of it out, but it's government-funded. That casino grosses a million and a half a day. Has for the last couple of years, since it took over the art gallery." He chuckled. "Gambling became more important than art. And 80 percent of those dollars are U.S. dollars. They're not Windsor, Ontario, dollars, they're coming from right there." He pointed across the river.

"So—Detroit got real jealous, real quick. So they're putting in *three* casinos. And the cheapest one is a billion dollars. The *cheapest* one is—not a million. A *billion* dollars." He shook his head in amazement. "But—the good news. GM bought the Renaissance Center. They're going to make a glassed atrium, and a greenway. The front of the Renaissance Center's on Jefferson Avenue. They're going to make that the back. The front door will now be on the river."

"On the river," I repeated.

He nodded. "They'll have a promenade between here and Stroh Place, which is just downstream of the bridge to Belle Isle. For the first time you're going to have substantial riverfront property opening up." He leaned on his forearms, his hands clasped. "There's still some huge issues that we're working on. I think toxics is a really, really, really important issue, and we have to continue to go down that road. We need to continue with control of contaminants at the source. We need to get on with contaminated sediment remediation, and we've got to keep the exotics out—we've got to do much more with urban sprawl, and habitat, and understanding biodiversity— I think we need a balanced approach. If we just go with the toxics agenda, we could lose all of this. You'd have a toxic-free environment, but if you didn't have the habitat, and no biodiversity—you know what I mean?"

"Yes."

"I think it sort of has to go together, Bill. And that's what the ecosystem approach is all about." He smiled. "Let me give you a greenfields story that is huge for Detroit right now.

"Detroit grew towards Lake St. Clair initially, and there are a lot of wealthy communities out there that are all called Grosse Pointe. Grosse Pointe Woods, Grosse Pointe Farms, Grosse Pointe Shores—they dominate the shoreline over there. And they're growing out towards the Clinton River, which is an Area of Concern, and a RAP. There's a beach, called Metropolitan Beach, down at the mouth of the river on Lake St. Clair. That area has the highest density of boats in the world. Nobody knows that. It's an *incredible* number of boats. And this beach has, like, three million annual tourists. Well, more people have moved out there, and the wastewater treatment plants there are smaller ones. They have a design capacity for the population density that was there when they were built, back in the late sixties and early seventies. We have now exceeded those design capacities. It takes less and less rainfall to have an overflow, which shunts waste out into the Clinton River and Lake St. Clair. So the frequency of combined sewer overflow events, and the frequency of beach closings, has gone up. There's also some failing septic tanks, which add to the problem." He shook his head. "Three million annual tourists swim at this beach. It's the premier beach out there. And periodically they have to tell you not to go into the water. Can you imagine the politics around that?

"And now you combine zebra mussels into that story. Zebra mussels come into Lake St. Clair, and allow the macrophytes to come back. They grow real tall in the water. And then a storm comes along, and they can't withstand the storm, so they break off. They float to the surface, and with the wave action and the wind action, they move inshore. And they create these *massive* walls of macrophytes. It's not a surface phenomenon, it's not just the top six inches of water—they start to get heavier, and they sink, and it actually forms a barrier to the movement of water from the Clinton River. So if you get a combined sewer overflow event, or you have too many failing septic tanks, you've got this *thing*—and it traps all the waste along the shore, and moves it into the bathing beach. And the people don't even realize that greenfields development and exotics are causing all that stuff. They think—well, somebody's not doing their job in the wastewater treatment plant." We all laughed. "Two years ago, they brought in the National Guard to pull all the weeds off the beach and put 'em in dump trucks. I think they spent $1.3 million. Two weeks later, it was all back. And that lake, as you know, falls through the cracks. Lake St. Clair is not an Area of Concern. There's no Lakewide Management Plan. So what are you gonna do? Does it have problems? Yes. Are you addressing them? Well—we're working on that."

We left the IJC offices and crossed back into the United States via the Detroit/Windsor tunnel. A bored customs official waved us through quickly.

"Citizenship?"

"U.S."

"What were you doing in Canada?"

"Research for a book."

"Have a nice day."

We drove to Belle Isle, Detroit's showcase park, which takes up all of the large island that splits the Detroit River just below Lake St. Clair. The park was large, and browned by dry weather, and nearly unpeopled. A few joggers; a few people fishing; a pair of rollerbladers. Several picnickers. There was a splendid little nature center with some of the best exhibits on Great Lakes natural history I've seen anywhere, but it was eerily empty: there were no people inside it at all, either staff or other visitors. A few of the exhibits were live, and these were rather sad: a blue racer trying to get out of its cage through a mirror, a turtle squeezed as tightly as it could manage into a corner of its small terrarium. Outside, the air was too thick for photographs, but small lakes rippled in green hollows and the Renaissance Center loomed impressively through the smog. Given good weather, it would probably be a very beautiful place—a Belle Isle in fact as well as name.

We traveled upriver to the Grosses. Along East Jefferson, the change was abrupt: ghettos, poverty, and black faces right up to the Grosse Pointe Park boundary; big houses, money, and white faces beyond. I was shamefully relieved to be where I wouldn't be mugged for having the wrong skin—shades of what people of color must deal with all the time. We had a late lunch in Grosse Pointe Farms and headed south to Taylor, where we had motel reservations. It was 4 P.M. Both the smell of the air and the Weather Channel suggested that rain was imminent.

Around 6:00 we ventured out again into the smog and impending raindrops to look for Humbug Marsh. We found it easily, a small island of wild on the border between Gibraltar and Trenton, bounded by chain-link fence, surrounded by decaying steel mills and lower-middle-class homes striving desperately to keep from becoming ghettos. Humbug in its natural state is extraordinarily valuable, even monetarily speaking—but there is no direct payoff to the local people, and you can't help but sympathize with the temptation offered by a $500 million development. Grosse Ile lies right out there across Trenton Channel, full of expensive houses, and the upscale development on the island is spilling over to the mainland. It's difficult to fault Gibraltar and Trenton for wanting a piece of the action. But they will lose far more, in the long run, than they will ever gain.

THE RAIN ARRIVED DURING THE NIGHT: not a downpour, but heavy enough to make the landscape thoroughly sodden. We splashed south along Interstate 75, accompanied by the tick and swish of the windshield wipers. Somewhere off to the left loomed the idea of Lake Erie, invisible but almost as present as if it could actually be seen.

At Monroe we turned off the freeway and headed east, toward the coast. Sterling State Park, Michigan's only state park on Lake Erie, loomed through the rain. A dimly seen silhouette behind a closed window waved us past the entrance kiosk. I stopped anyway and ran around to the tiny building's door. The attendant who answered my knock could have been a clone of the one at Hartwick Pines.

"Just go on in," he said.

"Could we have a map of the park? And maybe directions to the marsh trail?"

He gave me both, his expression clearly indicating that he thought we were out of our minds. But Sterling is another of western Lake Erie's numerous birding hot spots, and we didn't want to pass it without at least a glance at what it contained.

Erie is an anomaly among the Great Lakes. It is further south than the others, and older, and much, much shallower. The beds of the other Lakes gouge deeply into the region's bedrock; Erie more or less perches on top of it. Its average depth is only 58 feet. Though it has a considerably larger surface area than Lake Ontario—9,900 square miles to Ontario's 7,300—the "smaller" Lake to the northeast contains nearly four times as much water.

Not content merely to be different from the other Great Lakes, Erie also differs profoundly with itself. Lake Erie is actually three Lakes. The Eastern Basin, the part of the Lake that runs roughly from Erie, Pennsylvania, to the head of the Niagara River, has an average depth of around 100 feet and functions more or less as an oligotrophic body of water. The Central Basin, cut off from the Eastern Basin by an underwater ridge, averages roughly 60 feet in depth and is, functionally speaking, a mesotrophic lake. The Western Basin is marked off from the Central Basin by the line of islands stretching north from Sandusky, Ohio, to Point Pelee: it has a surface area of nearly 2,000 square miles but an average depth of only 28 feet, and its limnological classification is eutrophic. Bathymetric maps of the Western Basin's floor show a pair of straight parallel grooves running from Pelee Passage to the mouth of the Detroit River. These are ruts excavated in the bottom sediments by years of propwash from passing ships, whose hulls clear the bottom of the Western Basin by an average of less than three feet.

As is common with shallow bodies of water, much of the edge of Lake Erie is poorly defined. This is particularly true around the southwest corner of the Western Basin, where early settlers found an immense area of waterlogged forest so dank and mucky and dark they called it the Black Swamp. It was part of the bed of ancient Lake Whittlesey, a forerunner of Lake Erie, and it was so wet that the Lake didn't really seem to have gone away but only into hiding. The settlers rousted it out and expelled it. The forest was cut down and dried out. Deep in its damp heart, near the mouth of the Maumee River, a town was platted and given the hopeful name "Manhattan": it grew into Toledo. Around it spread fields and factories. The Black Swamp was reduced to fragmented remnants.

But there is this thing about wetlands: they are, acre for acre, among the most productive ecosystems on the planet. They are nurseries of aquatic life, protecting the young of numerous species, including many which spend most of their lives in deep water. They filter and polish the water that flows through them, removing impurities and turbidity and improving oxygenation. As much as humans seem to hate them, birds adore them. This is true not just of waterfowl, but of songbirds as well. Swamps and marshes are full of things to eat: insects, fruits, grains, small fishes. The trees provide nesting sites; the damp ground discourages terrestrial predators. If you plot the routes of the three great bird-migration paths of North America— the Eastern, Central, and Western Flyways—you will see that they pass directly over the cream of the continent's wetlands. This is no accident. The birds know precisely where they are going.

So as the size of the Black Swamp shrank through the nineteenth and early twentieth centuries, the value of the remaining disconnected bits of it followed the classic law of supply and demand, and rose. Today, most of what is left is protected as wildlife habitat, a string of preserves stretching south and east around the corner of the Lake from Monroe to Sandusky. Sterling State Park anchors the northwest end of this string. Rain or no rain, a visit was required, despite what it might cause the gate attendant to think of the mental acuity of middle-aged couples with Oregon license plates.

Actually, aside from its dampening effect (literally) on photography, the rain was probably an advantage. The defining fact about wetlands is, after all, that they are wet. Most of Sterling's marshes are actually reconstructed—the park is built around four artificial lagoons fed by canals from Mason Run, a tributary of the River Raisin— but although that has a significant effect on their function as part of Lake Erie's ecosystem it does not greatly alter their use by birds. Shrugging into ponchos, we walked the mile-long trail around the biggest and marshiest of the lagoons. Herons and egrets stiltwalked the shallows; coots paddled lazily through patches of open water. Marsh mallows, whose roots once formed the raw material for the confection of the same name, raised pink crepe-paper petals along the banks.

At its farthest reach from the parking lot, the trail traversed a berm between the lagoon and the principal feeder canal from Mason Run. On the far side of the canal, twin smokestacks lifted through the slanting rain. They belonged to the Ford Motor Company's Monroe Stamping Plant, which represents two things. It is one of the Great Lakes' most successful pollution-cleanup stories. And it is an excellent demonstration of the myopia we fall into when we define pollution-cleanup success.

Ford opened its Monroe plant in 1949, on a bit of reclaimed Black Swamp just north of the mouth of the River Raisin. The plant made automotive body parts, and in addition to stamping them—that is, cutting and shaping them with stamping mills—it painted and plated them. The wastes (remember, this was 1949) were discharged directly into the river. By the seventies, when direct discharge came to an end, the river's sediments, water, and aquatic organisms were heavily contaminated with oil and grease, volatile organic compounds (VOCs), a variety of heavy metals—and up to 40,000 parts per million of PCBs. In 1982, the River Raisin became one of the IJC's Areas of Concern.

I want to emphasize, here, that we are talking about a success. Ford is one of the good guys. They stopped dumping voluntarily in the early seventies, when the problem became apparent, and they cooperated in the cleanup process. By October 1997, $6 million had been spent—much of it supplied by Ford—and the contaminated sediments were gone. Tests found PCB concentrations reduced from 40,000 ppm to less than ten. The River Raisin AOC was not yet delisted as we walked the berm that day in the rain, but it was definitely well on its way.

But I mentioned myopia. To see it, look again at that last paragraph. It mentions PCBs. It does not mention VOCs, oil and grease, or heavy metals. Cleaning up the PCB-contaminated sediments may well have cleaned up these other things as well. But no current report of the Raisin River cleanup that I have managed to find will tell you anything about them.

Do we wish to clean up the Lakes? Then let us do so. I strongly approve of the effort. PCBs are part of what needs to come out of the water and the sediments and the biota, and I approve of that, too. But let us not be shortsighted about it. The proper test of a clean Lake is not whether the PCBs are gone. It is whether the Lake is clean.

We made one more stop that rainy morning, at Ohio's Crane Creek State Park east of Toledo. No artificial berm separates Crane Creek's swamps and marshes from the Lake: unlike the lagoons at Sterling, they still function as nature intended. This is one of the major gathering areas for migrating birds crossing Lake Erie—a south shore equivalent of Point Pelee—and like Pelee, it is highly regarded by birders. We hiked the half-mile boardwalk through the swamp forest, and I'm sorry to say I was not particularly impressed. There had been more bird life back at Sterling. After the boardwalk, however, we walked down to the beach, and there were zebra mussel shells. Windrows and heaps and truckloads of zebra mussel shells. They were piled a foot deep along the edge of the water, and as they rolled in and out with the breakers they tinkled like toy money. Gulls perched in a long row on a corrugated metal breakwater, and the rain came down, and waves came booming in under the big gray sky and instead of crashing they tinkled. Lake Erie residents are probably used to it, but it was one of the odder natural sound effects I can remember.

We stopped for lunch in Port Clinton at a Subway that hadn't been there in 1983, then turned left at a Wal-Mart that hadn't been there either and headed up Catawba Island. Separated from the Ohio mainland only by a narrow slough which is closed at its western end by a sandbar big enough to have homes and roads built on it, Catawba is less an island than a gateway to other islands. Fifteen minutes to the north across the flat, restless surface of Lake Erie lies one of the

A drift of zebra mussel shells on a Lake Erie beach at Crane Creek State Park, Ohio

loveliest archipelagos in the United States, or perhaps on the planet: the Bass Islands.

The Bass Islands are low and flat and sculpted by deep bays and inlets. They are made of dolomite, and though the ice tried—glacial striations as much as six feet deep show how hard the ice tried—it could not wear them down. There are five major islands: South Bass, Middle Bass, North Bass, and Kelleys in the United States, plus Pelee in Canada. The sea around them is spangled with islets—Starve, Green, Ballast, Sugar, Rattlesnake. Put-in-Bay, on South Bass Island, sheltered Commodore Oliver Hazard Perry's fleet on the night of September 9, 1813, just before the Battle of Lake Erie. Today the bay shelters large numbers of pleasure craft. This is Ohio's Riviera, or Ohio's Greek Isles, or perhaps both. The islands are crowded with summer homes and guest cottages, and they throng with vacationers. Sun dances on water and deep, wet woods; coves and caves beckon.

Sailboat masts stand in forests along the docks. There is a thriving wine industry.

On Gibraltar Island, in the middle of Put-in-Bay, a nineteenth-century railroad tycoon named Jay Cooke built a small castle. Today it is part of Stone Lab—the Franz Theodore Stone Laboratory of Ohio State University, the oldest freshwater biology research facility in the United States. In 1983 I had toured the island and circled it in a small boat piloted by Stone Lab's director, Dr. Charles E. Herdendorf III ("Eddie" to friends and family, a fine name for one who studies water). Inquiries in the spring of 1998 found him retired and living in Huron, on the mainland, but still teaching a couple of summer courses at the lab. Eddie and his wife, Ricki, maintained a cottage on the north shore of South Bass Island, nestled in trees at the edge of the water, where they lived while Eddie was teaching. The cottage had a spare room. Would we like to spend the night?

At the tip of Catawba we caught the ferry to South Bass. It looked about the same as the one Larry Chitwood and I had taken in 1983, and may indeed have been the same boat; but where then it was only about three-fourths full, this time it was crammed. If we had been driving a bigger car we wouldn't have made it; but I squeezed the Escort on diagonally just in front of the boat's rear ramp, and by not raising the ramp all the way the crew managed to clear the back bumper by an inch or two. We rocked over the Lake Erie swells in the thick gray air, the island rising out of the mists before us like a legend of dragons. Safely ashore on the land in the middle of the water, we cruised slowly into the village of Put-in-Bay, dodging golf carts and bicycles, the black and white cruiser of the local constable directly behind us. In the round bay Gibraltar Island loomed like a small version of its namesake rock. We circled the water to the northwest and found Stone Lab's onshore facility, housed in a white frame two-story cottage. I told the receptionist I was looking for Ed Herdendorf. "Right here," said a voice from the copy machine around the corner, and when I looked, Eddie was there, looking not a day older than the last time I saw him.

With Eddie and Ricki we caught the "ferry"—an open aluminum boat with an outboard motor—to Gibraltar. "We usually eat with the students," Ricki explained. "It's part of the deal Eddie worked out with the university when they asked him to keep teaching." Ricki was at least twenty years younger than her husband; her thick mane of hair tumbled down her back in tawny ringlets. At the boat's tiller was John Hageman, Stone Lab's physical manager: dark-haired, laconic, and somewhat broad of beam, he was dressed in a baseball cap and a faded orange T-shirt and looked like a truck driver on a fishing holiday. Graduate student Melissa Haltuch, a slim, attractive twenty-something, completed the passenger list. The boat moved through masses of dark floating plant matter, bits of macrophytes—rooted aquatic plants—sheared off by wave action and hull passage. They covered much of the surface of the harbor.

"I wanted you to see this," Eddie told me, gesturing at the flotsam of macrophytes. "Those weren't here in '83—the water wasn't clear enough. It's a *lot* clearer now. When the light is right, you can see bottom all the way across to Gibraltar."

"Would that be affecting the yellow perch?" I asked.

Eddie looked puzzled. I told him about the concern I'd heard regarding plunging perch numbers. "But they're not declining," he protested. "Not in Lake Erie. You can't manage a lake for walleye and yellow perch simultaneously, and the walleye are up, so the perch are probably down a little. But they're still plentiful."

John maneuvered the boat skillfully up to Gibraltar's dock and we walked up the hill to the dining hall. Eddie looked at the chalkboard menu and nudged me. "You were worried about yellow perch," he said. "We did this just for you." Moments later my plate was being filled with breaded, deep-fried yellow perch. John Hartig had quoted a price of ten dollars a pound for yellow perch in Windsor. This perch had been pulled from Lake Erie, not purchased, but it was still difficult to convince myself—even as I went back for seconds—that I was not eating my hosts out of house and home.

IT WAS RAINING AGAIN, HARD, WHEN WE got back to the Herdendorfs' cottage, and the dark had come down as only dark on an island can. The cottage was tiny and nestled in what looked like a natural woodland. "There was a lawn around it when I bought it in 1969," said Eddie. "I just let it go. The neighbors complained about my ragged lawn, but it's past that, now." Inside it was cheerily bright. Lighthouse pictures lined the walls. There was a mutter of nearby surf.

Eddie and Ricki had invited several people over for the evening, and around 7:00 they began to trickle in. John Crites—short, white-haired, and slightly stooped, an emeritus professor of zoology from Ohio State—showed up first, with his wife, Phyllis. Melissa Haltuch and John Hageman arrived together a few minutes later. Dave Moore, a quietly intense biologist from Utica College in New York who was at Stone Lab for the summer, rounded out the party. Eddie broke out a bottle of Bass Islands wine. Without asking, he tossed a bottle of beer to John Hageman. Evidently they had been through this little ritual numerous times before.

"So what would you like to ask us?" Eddie asked me.

"Why don't we start," I suggested, "with what you think is the single most important change that has taken place in the past fifteen years."

There was a short pause which was not a silence because of the rain on the roof and the waves hissing like cats at the black dog of the shore.

"I think the most important change—," John Crites began. "I've been around Lake Erie since 1956. I think the most important change is that the Lake's been cleaned up. The chemical situation that we had—phosphorus, and nitrogen, and so forth—has cleared up a great deal. You can see the results in the Lake itself, and it's not always good things. Time was when I could stand on the rocks in back of my house and catch white bass like everything. I can no longer do that. But on the whole, that's the biggest change I've seen. And I don't credit the zebra mussel for doing it, either. I think that the work on point-source reduction of materials coming into the Lake, and reducing the chemicals and the nitrogen and phosphorus, has had more effect than the zebra mussels. But there are a lot of people who disagree with that."

"Yeah," agreed Eddie. "We began to see improvements to the Lake *before* the zebra mussel invasion. I think it's probably a mixed cause. The zebra mussel came along, and we saw an incremental jump, but we certainly were on the road to recovery before that time."

"I've seen a paper in the *Journal of Great Lakes Research*," I said, "that coordinated Secchi depth in Lake Erie to rise and fall of zebra mussel populations. So there has got to be *something* to that." Secchi depth is measured by lowering a black and white disk the size of a dinner plate into the water until it disappears, raising it until it comes into view again, and averaging the two readings. It was under four meters in Lake Erie in 1988, when the zebra mussel first appeared. Today it is over seven meters. Much of that near-doubling appears to be correlated to the spread of the mussels.

"But I think the Lake was on its way to recovery before the zebra mussels," John Crites repeated. "There may have been an acceleration as far as clarity of water is concerned, but I think the chemical condition of the Lake was on its way before they were introduced."

"And we've changed the population of phytoplankton," said Eddie. "I can recall in the sixties taking a trip across the Lake from Sandusky to Canada, and there was actually a green wake behind the boat the entire trip. That was gone in the late seventies and early eighties. We had turned the corner on that situation. The importance of that is, the blue-greens are not particularly good food for zebra mussels. I think diatoms are more a preferred food. So the Lake had to improve to a state in which diatoms were again the dominant phytoplankton before the zebra mussels had the appropriate food."

"So the zebra mussels couldn't invade *until* the Lake was clean?" I asked.

"Well—they didn't, anyway," Eddie said, with scientific caution. "And it certainly is a much more favorable situation for them with a more moderate phytoplankton."

"In your book," Dave Moore said to Eddie, "in the *Lake Erie and Lake St. Clair Handbook*, you had a table that has the Secchi disk readings.

And they show—well, it must be a meter or more of improvement in water clarity. In the late eighties it just *jumped*."

Eddie smiled. "Dave has been studying the higher aquatic plants," he said. He looked at the Utica professor. "Why don't you tell us what you've seen?" he suggested.

"When I was a grad student out here," said Dave, "it was dominated by turbidity-tolerant types. And when I came back in '85 to teach, there was a big difference. You could see there was a change going on. And then I came back in '94, and it was a world apart. Completely different. Things that had been common were being displaced, and we were seeing other types of plants come in. I've watched the expansion of pondweed, for example. Just *huge* colonies out in the middle of Put-in-Bay Harbor, and they're expanding."

"Would that have anything to do with the increased clarity of the water?" I asked.

"Uhm-hmm," he nodded. "And a change in substrate just a bit, from a highly organic to a bit sandier bottom."

"OK," I said. "My understanding is that zebra mussels change the character of the substrate."

"Yeah," Dave agreed. "I've actually looked at it over a couple of years. And it has changed. What was very organic is now sandy, which reflects the light. So the amount of light has changed, which allows the oligochaetes to do a little better."

"We're ending up with a sort of a confounding situation," Eddie remarked, "in that we had improvement coming along, and we have the zebra mussel—it's hard to apportion what the total improvement was. Or the total—*change,* I don't want to say this is an improvement, because John—" He turned to John Crites. "Why *aren't* you getting white bass anymore, John?"

"I don't know," said John gloomily. "I really don't know."

"We used to see them," Eddie said, looking a little dreamy. "They would be coming up, driving the shiners up to the surface. And you'd see them just swarming, and then you'd see the birds coming down, the gulls, hitting the shiners from above, and that's where you would catch them."

"Last year was a pretty decent haul," John Hageman put in. "For about three weeks, right in the harbor mouth, we got white bass, and—"

"This year?" asked John Crites anxiously. "Did you see any this year?"

"This year there've been some close to Rattlesnake Island, a couple of schools—we were out a week ago Friday."

"What about white perch?"

"White perch seem to be way down," said Hageman.

"I don't know if it's true or not," said Crites, "but people were saying that the competition with white perch was the reason that white bass were down. I don't know if that's true."

"The mayflies are back," I said, "and some people are citing that as evidence that the zebra mussel has converted Lake Erie from a mesotrophic to an oligotrophic system."

There was a short silence while this sank in.

"From a mesotrophic to an oligotrophic system," Eddie said finally.

"Yeah," I said.

"Bit of an exaggeration," said John Hageman.

"I find it very strange, personally," I agreed. "But there was an article in the *Journal of Great Lakes Research* that said that, and one of the evidences given was the increase in mayfly populations. The return of the mayflies."

"Well, I don't know," said Eddie. "To me, mayflies are not a particularly good criterion when you're talking about trophic status. If you look at the mayfly, it's traditionally been thought that it was an anoxic situation that caused its demise. And that may be partially true. But 1955 was also the peak year of production of DDT in this country, and that's the year they disappeared. And they disappeared from a lot of areas that weren't subject to anoxia. They went *every*where. So it always has bothered me that anoxia was pegged as the only reason for the loss. Certainly that was a stress on the system, but I think it was the organic pesticides that did them in."

"Well, that would be a distinct possibility," Dave Moore said.

"I'm surprised they came back," Eddie continued, "because once the chironomids moved in I thought maybe they would establish themselves so well that the mayflies would have trouble." He looked around the room. "Have you ever noticed—when you see reports of the numbers of mayflies, there are two big zeros right here off the island? That seems troublesome. They're traditionally areas which had high populations."

"I was going to say," remarked John Crites, "of course, things never get back."

"No," Eddie agreed.

"The conditions now are not the conditions we had in the fifties," Crites said. "The water clarity we saw in the fifties wasn't nearly as clear as it is now. And there were *more* mayflies. That's not to say they aren't coming back. But they're nothing like they were."

"When I decided to ask people about change," I said, "I expected that the main thing mentioned would be nonindigenous species. But actually, only one person has mentioned that, before tonight. Most people have said that the major change has been shoreline sprawl."

"Well, there are many, many more folks on the Lake than there were," said Crites. "You can look out here mornings and see boats stretched from Green Island to Rattlesnake, and on out beyond. On some Saturdays, you can see—is it fair to say, hundreds of boats?"

"Easily," agreed John Hageman.

"And this was not true before. That's got to have some kind of effect—the number of boats, and the population. I don't know if it's been measured at all."

"What's happened," explained Eddie, "is that most of the Lake Erie shoreline, if it was subject to flooding or erosion, has been rocked over with limestone riprap. There's hardly any of the natural banks left. There's an issue that John and I are interested in with coastal wetlands. Coastal wetlands are very important to the health of the Lake—they're cycling places, they're places of water storage, and they're places of fish spawning. The DNR claims to be restoring wetlands, and doing a wonderful job. And they did an excellent report. However, what they're doing is diking off the shoreline and creating a wetland behind it. I've accused them of creating duck ponds, and nothing else. Because these wetlands do not function as coastal wetlands—although they claim that they do, and they use the rationales that we've written about why we need them to support getting money to do what they're doing, which is just diking off the shores."

"There's no connection to the Lake," I clarified.

"That's right," agreed Eddie. "There's no connection to the Lake. All it does is allow a place for migrating waterfowl to land, which is important, but it's just one cause. And actually they're taking fairly functioning wetlands now, and diking them off, and sealing them from the Lake."

"Aren't they controlling them now?" Ricki asked. "I've heard recently that they're controlling the levels in them, they're not completely diking them off all the time. They're going to be able to manage when they want to open it up, or when they want to close sections of it."

"What they do," explained Eddie, "is, they *do* control the water level—but primarily they draw the water down to grow plants that will attract ducks. There's only one that they've experimented with, on a minor scale, to have a fish opening. The reason they don't want to do this is they don't want to invite some of the fish, like carp, that will cause turbidity and uproot plants. But that seems to be where the funding is to do these things—from Ducks Unlimited, and people like that, who have a very narrow focus on what they want to do. They aren't really considering the Lake as an ecosystem. That's what I complained about. Right now, in this supposed time of enlightened management, there's no enlightenment whatsoever."

"Well," I said, "that sort of segues another thing that people have talked about, which is changes in government. The most prominent example of that, of course, is the RAP process—almost all funding has been withdrawn by almost all agencies around the Lakes. It's pretty much back to the local citizens. And when local citizens are the ones responsible, then the ones beating the drum for something are the ones who raise the money and get their item done."

"I don't know if I quite agree about the fact that the money is being withdrawn," said Eddie thoughtfully. "The whole LaMP Process is really a coordination effort of government agencies, so they're putting funding into at least the planning process. And the RAP process in Ohio is managed through the EPA. And other than the government serving a role as encouragement, I don't know that government funds are appropriate. I'm not a big government person."

"It's more of a marketing problem, actually," said Ricki. "People are not buying as many improvements for the Lakes."

"How do people feel about the contaminants issue?" I asked. "Is it really improving, or is that simply window dressing?"

"Well, I think there's an obvious thing you can see." said Eddie. "At most of our major harbors—Toledo, Lorain, Cleveland—Huron, even—the harbor dredgings were to be placed in a confined disposal site. They were built right at the harbor entrances, and they were projected to last maybe twenty years. Well, the Huron one has been there *over* twenty years, and it's not even half full yet. I haven't seen them put any dredgings in there at all in recent years. It sits there half full—makes a nice wetland—" He grinned. "But I think that's something that the public sees. It sees these things not filling up."

"A question that has occurred to me," I said, "is how long-term a confined-disposal facility really is. Because it seems to me what you've done is just take the contaminated material and concentrate it in one spot. You haven't really solved the problem."

"You hope it's not going to leak," said Eddie. "But—"

"Is there anything else on the horizon?"

"No. Again, the premise was, that with point-source controls—which are the obvious main source of toxic metals and organics—they should be controlled to the point where the sediments will no longer be contaminated. I think that's the whole idea."

"But what about nonpoint, Ed?" I persisted.

"Well, a lot of it, like the fungicides used in farms, has a shorter lifespan in terms of sticking around the environment. We heard a presentation recently that said, even though we've had 80 percent volunteer non-till farming in some areas, it's not caused a like cleaning of the rivers in terms of the sediment load. Sediment's not toxic, it's not a contaminant, it just means that the rivers have as much of a load—or nearly as much—as they had before all this non-till farming. It was very disturbing when we heard those numbers. It's hard to understand if the data's right, or what's wrong, or what's going on. But one of the things that I think is real important is having greenbelts along the rivers, not just non-till. Because farming still goes right up to the edge, so you're still getting runoff of sediment right into the streams. A buffer zone, a greenbelt, seems to be the way to go. But when you're not providing any compensation to the farmer, and they're taking land out of production, it's easy to understand their reluctance. Getting back to that point about dredging, though, I don't favor that either. Don't disturb it. Let it cover over with cleaner sediments, and let that just be a sink."

"That's one of the things the environmental community has been pushing," I reminded him. "To just let it naturally seal. The problem with that from the standpoint of the users is that you *can't* dredge. So you get pleasure-boat harbors that fill up, and you can't put the boats in them anymore."

"That's right," said Eddie, "but my feeling always has been that when you get away from the tight harbor areas, the concentrations are not such that one would have a concern. I think that the dispersal is such that, when you move much away from the harbors, it's just not there. It's not a problem." He grinned. "All of this is speculation," he said. "It's great to be retired, and speculate."

"It will affect the benthic flora," Dave Moore pointed out.

"Yeah, certainly the bioturbation might extend down farther," agreed Eddie. "Yeah, it's true. With the mayflies coming in now. But again, I don't think it's a problem. When you look at the kinds of contaminants coming in, I think they're dispersed so widely when you get to the open Lake that I don't think you *have* sediments that are contaminated."

"Let's return to the question of sprawl for a moment," I suggested. "Because when you're talking about nonpoint sources, one of the major problems *is* the sprawl. Increased development along the shoreline increases the amount of pavement. Which increases the runoff. Which then increases the loading to the Lake."

"We have a law in Ohio," observed Eddie, "that every time you build a development, and you take up a certain amount of surface acreage with pavement or buildings, you have to replace that with a retention pond. Or a *de*tention pond, actually. It has to be maintained as a sterile kind of pit. Those are theoretically supposed to hold these contaminants. The primary purpose is probably for flood control, but I think they also serve as a sink for sediments."

"The idea is to hold the runoff," I clarified.

"Right. It's to take the place of the land that has been paved over. And it's to give an infiltration point, so the groundwater can infiltrate back in. And also a sludge-detention area."

"OK. The immediate thought that comes to my mind with something like that is that it also is concentrating."

"Oh, it certainly is," he agreed. "It's a much smaller area. I don't *like* them. But it's the best the engineering community's come up with so far."

"How has this affected the Lake?"

"Well, actually, I think it's impacted the tributaries coming into the Lake more than the Lake itself. I can remember streams that seemed to flow a good deal of the time, because the groundwater would seep in, and would slowly be released to the tributaries. Now, with water running off so rapidly, there is no continuity in terms of water coming into the tributaries. So we have streams that are either in flood or dry. I say that just as an observation—it's happened in the last several decades."

"Definitely," agreed John Hageman. "And more prevalent flooding that happens when we *do* get rain, and it doesn't seem to be diminishing at all. It's accelerating. Instead of having a ten-year flood every ten years, you have it every *two* years. Instead of having a hundred-year flood every hundred years, you have it every *ten* years. And what are we going to do about it? Well—don't worry, next year they'll dig 'em deeper, and straighten 'em out, and everything'll be OK." There was laughter. John, who must be hell to play poker with, didn't crack a smile. "Yeah, I just went through that fight last fall over on the mainland," he said. "The middle branch of the Portage River hadn't been 'maintained' in about thirty-five years. And the county engineer had this great idea that he was going to cut down thirty-five years' worth of forests, on both sides of the river, for a four-mile stretch. That goes against the grain of every conservation agency, every state water conservation district—the Division of Wildlife's efforts—the list of partners is close to eighty people on this pamphlet that I gave him to look over. Their goal actually is, within the next eight or ten years, to establish a couple of million acres of riparian corridors. And I handed him this, and I said, 'Well, here's four less miles that they're gonna have.' "

"What was the rationale?"

"There's farmers along that stretch of river that are being flooded several times a year. Four years ago they were only flooded once every *several* years. And the state's gotta do something about it, is their cry. You know, 'We're paying ditch maintenance fees—get in there and *tame* this river.' But fortunately—this is on the outskirts of Bowling Green—most of the people who now live along that stretch of the river are people who moved out to the river *for* the riparian. And if it floods every so often, well, that's just the price of being on a floodplain. Most of them have the right attitude about it."

"The landowners actually said that," marveled Melissa Haltuch. "That's pretty impressive."

"But isn't that a long-term land-use problem?" asked Dave Moore.

"It's a long-term land-use problem," John agreed, "because they've allowed too much development along the river—everybody wants a river view. And all the farmers have to get their water off the fields quicker, and it adds to the immediate flow during a rain event."

"It goes back, too, to the diked wetlands," observed Dave. "Because farming is going right up to the diked wetlands, so there's no place for the wetlands to migrate. So pretty soon you don't have a diked wetland there, there *is* no wetland, and then the Lake comes up and floods the farmland."

"They build roads that have to be where they're at," Hageman agreed, "because that's where they built 'em, and they're gonna keep 'em there, come hell or high water—literally. And so if the wetland tries to migrate inward they put the rocks up higher. And as the wetland recedes, and Lake levels go down, people scurry to put houses along the shoreline, only to be flooded out when the Lake comes back up again."

"Have the Lake levels been following the same cycle, over the last

fifteen years, that they followed for the previous time of record?"
I asked.

"They continue to slowly rise, it appears, in a long-term way,"
said Eddie. "I think we're running well above average. Of course,
the average goes back and includes all of the dustbowl years, in the
thirties, in which the rainfall of the whole country was so depressed
that it skews the average down."

"Hasn't there been an overall tendency since the eighties to come
up?" asked Dave Moore.

"Yes, it has," Eddie agreed. "We continue to rise slowly, and stay
up there. So it looks like we're way above average. But if you discount
that, we're just kind of bouncing around a plateau."

John Hageman looked serious. "The Niagara River is rebounding,
so we're gonna get deeper before we get shallower. Over the course
of the next couple million years."

"Well—I'm not sure that could explain it all, John," smiled Eddie.
"But just to show you how attitudes change—we were talking about
urban sprawl and things of that nature, and we talked about the
shoreline now being all rocked over and protected. But when you
look back—the Beach Erosion Board of the Corps studied erosion
in Erie and Lorain Counties in about 1945. They had a strategy for
erosion control. And what it was, for certain areas that were just
farmland, was to let them erode. Because that was supplying sand
to the littoral system, and nourishing beaches on downstream. So all
the area between Vermillion and Huron was to be left to erode, so
that Cedar Point would be nourished." He grinned. "Of course, that
wouldn't go over so well with the people living in that area now, in
the condo that I live in."

"With the increased development that's going on along the shoreline," I said, "there's no longer any sacrifice areas. And there seem
to be very few places where we've been so far where the shoreline is
still open."

"Well," Eddie began, "with the same—"

"Is there something *wrong* with people living along the lake?" Ricki
exploded. "I think it's great!"

"No," I said. "It's not wrong. But there's something called the
Fallacy of Composition at work. Many people doing things that are
perfectly reasonable, and logical, and beneficial to themselves, can
end up with an unreasonable, illogical, *bad* situation in the long run.
When you build up *all* the shorelines with development, then you've
lost something."

"I don't know," said Ricki. "We heard a talk, and I can't remember where, but the speaker said Lake Erie belongs to everybody in Ohio, and the entire shoreline—"

"He proclaimed the DNR's goal was, I've forgotten how many years hence, but to own 50 percent of the shoreline again," added Eddie.

"So how many plots do we need?" asked Ricki.

"Yeah," agreed Eddie. "It's troubling, because there's a lot of farm-land being converted to residential sites. You're right. Individually, it's just one little plot of land. But collectively, you begin to get huge acreage converted."

"The classic example is right along Catawba Island," said John Hageman. "One person's half-acre house that they fill in and make a driveway over the top of the wetlands has virtually no effect. But they have house after house after house after house, and pretty soon the barrier beach wetland is just a remnant. That's the case over there. Terrible flooding in the spring, with the Lake being high, and a couple of sustained northeast winds that just deluge all those residences over there. So now they're all lifting their houses, and fortifying, and digging in."

"Yeah," agreed Eddie, "they can't see the Lake in some cases. They've put the mounds up so high, and the houses are sitting down low that you can't even see the Lake over the top of the rock."

John frowned at his beer. "Instead of abandoning ship, they're digging in for the long haul. It's disheartening. I thought the Lake was gonna win a couple of those battles."

"Some of these are barrier beaches," explained Eddie, "just like the barrier islands along the East Coast. They tend to migrate landward over time. And when you've got a house, and you're trying to hold that position at all costs—you end up defeating the beauty you've come to enjoy." He looked at Melissa Haltuch. "I'm going to ask Melissa a question," he said. "Melissa's doing research on zebra mussels, and predicting where they're going in the future, in terms of numbers. Her model predicts mainly on sediment type. But she's done predictions through the year—2000?"

"Right," said Melissa.

"Can you give us a number, based on current populations, what they might be like at the millennium?"

She thought a moment. "I'd expect the population to expand to cover 60 percent of the area. Approximately 60 percent."

"What are they doing now? What percentage are we at?"

"Probably around 40 to 45."

"And some of the areas that actually have no populations now, are predicted to—"

"Have about 20 percent coverage," Melissa finished for him.

"And those are much bigger areas," said Eddie. "So lakewide, might we see 30—or maybe 50, 100 percent more zebra mussels in Lake Erie?"

"Well, if you look at my model, you could say that," she said, "but the model predicts only on sediment type, and absolutely ignores other factors."

"What I'm trying to get around to, is asking"—he grinned—"this is your thesis defense."

Melissa laughed. "Sure," she said.

"Your model says that there will be a significantly greater number of zebra mussels in the future. Maybe 50 percent more, or maybe doubling. What do you think that might mean to the ecology of the Lake?"

"I think you're going to see softer substrates shifting to harder substrates," she said. "And they find smallmouth bass now in association with zebra mussel reefs, so you're going to see changes in smallmouth bass populations."

"Smallmouth might become one of our more dominant fish species," Eddie suggested, "rather than localized in the island area."

"Right. And you also find round gobies in association with zebra mussel reefs. In the remote vehicle [ROV] surveys that I've done this summer, wherever there's zebra mussels, there's gobies. Even on soft bottoms."

"Everybody's talking about the zebra mussel," I observed. "Nobody is talking much about the other non-native species that have invaded. There's 139 of them in the Lakes at last count. What problems do some of these others cause? Let's start with the goby. What does it do to the ecosystem besides simply be there?"

"I'm not a round goby expert," Melissa began, "but I think that it would compete with other planktivorous fish for resources."

"The jury's still out," said John Hageman. "Other people who have watched em claim that they're an aggressive fish, very territorial—they like to set up a little territory, and defend it against others of their own kind, and other species. And it may cause the reduction of some of our native fish, darters primarily."

"We've noticed with the ROV," said Melissa, "that the gobies don't really react to it. We can set it down right next to them, and they don't react more than a little bit. We can follow them around real easily. Some of the others—as soon as you start moving, it makes you kind of a predator, apparently."

"We have a scud," said Hageman. "An amphipod called *Echinogammarus,* which is a European species as well. And this *Echinogammarus* has actually become the dominant scud out in the Lake, especially around zebra mussel reefs. It's very abundant, it provides a lot of food for yellow perch—and the whole question comes up again. Zebra mussels are in contaminated areas. Do they pass those contaminants to the scuds? Do the scuds pass them on to the yellow perch? Does it end there?"

"One school of thought I've seen on that," I said, "is that zebra mussels aren't very good concentrators. What they tend to do is to eject toxics in their pseudofeces."

"That may be true," agreed Hageman. "But some of the food that the scuds feed on *is* the pseudofeces of the zebra mussel."

"So they're pulling those things out of the water at a higher rate," explained Melissa, "and making them more available."

"OK," I said. "But I notice we've got right back to zebra mussels."

"Oh, we've been surrounded with 'em since 1988," said Hageman. "Really, zebra mussels have helped put our facility on the map again. With the number of research projects that are centered in this location, we're sort of—brainwashed that zebra mussels are a major factor in the changes that we're seeing. Not *the* major factor, or the only one, but certainly a major player."

"I was going to say," said Eddie, "that if you would be on a dive with us when we snorkel around the island, you would certainly understand *why* they're so pervasive in our thinking. But I had a question for John Crites. John's specialty is fish parasites. And he has looked at them over the years, and a number of his students have worked here." He looked at John. "Have you noticed any changes in the parasitology?"

John demurred. "I haven't worked parasites in Lake Erie for twelve years. I retired ten years ago, and I've been working in the ocean ever since."

"Well, I was just wondering if any of your colleagues have been mentioning anything—"

"No," said John. "In fact, if I were active now, my guess—it's a *guess*—is that it's quite different. The whole system is based on the invertebrates. If you change the kind of copepods, you're changing the scuds that are out there. You change the number of mayflies, and the invertebrate intermediate hosts—the transfer hosts for the parasites—and you have a quite different situation. But I *don't know.* If there was someone in parasitology active in Lake Erie now, I would suggest that they could get a lot of money. Because the data's here, from back to the forties and thirties up through the eighties. And it would be really nice, as far as I'm concerned, to know what's out there now. But I really have no idea."

"Is it likely that the zebra mussel fits in any way?" Eddie asked. "Do any of those go through zebra mussels, or—"

"No, that's been checked. They're not particularly good intermediate hosts. But what they did was take out most of the big beds of unionid mussels that were around. And *those* were intermediate hosts for a number of things."

"That's right," breathed Eddie.

"And they're gone. So I assume that all the parasites are gone. In the fish. But I *don't know.* Because nobody's studied it, or checked for these things."

"They were fish parasites themselves," Eddie said. "They all selected—each one a different fish host, didn't they, the unionids themselves?"

"The whole ecology has changed," said John Crites.

"Yeah."

"And at my age, I'm not gonna *do* it." We all laughed. "I'd like to, but I can't. I've suggested to some people—there's a fellow up at Michigan State who's working in Saginaw Bay. Places like that. This is something that ought to be checked."

"One thing that may have an impact on the Lakes," I said, "that seems unpredictable, is global warming. Has there been anything that you could pin to that yet?"

"You know," said Eddie, "they took the temperature gauge out of the hatchery, that had all that historic record back to 1919, when they converted it to a visitor center. The recording gauge is now history. If they put it back in, we could begin to answer that question."

"A few years ago, the whole Lake froze for the first time in fifty years," said Hageman. "So we know it got cold that year. But then the next year, it didn't freeze, for the first time in a decade. It *didn't* freeze. So—"

"Well, last year it was warm," said Phyllis.

"Have you heard any predictions for what global warming will do to the Great Lakes?" I asked.

Eddie nodded. "Oh, yes. There was a conference in Chicago a few years ago. And depending on which presenter you had there, it was either going to become wetter or dryer or hotter or colder." He smiled, and the rest of us laughed. "It's really all *over* the map, in terms of the kind of predictions we have. I left the conference thinking there was no way you could think about it. Have you heard anything credible?"

"Actually," I said, "I have not. I think the general sense is that there will be greater rainfall."

"What makes me not so concerned about it," said Eddie, "is—looking back at some of the global change literature, and tree ring dating, and some of the polar ice cap temperature dating, and things of this nature—when we go back to the Middle Ages, it was apparently a fairly warm time. Maybe four degrees warmer than it is now, on the average. This is when Iceland and Greenland were settled, and it must have been a very pleasant time in England. We had *quite* a difference in regime. And then we went back to a colder period, and apparently we're now climbing out of it. Back in the 1800s it was still quite cold. So when you look at it in a broader time frame, it doesn't concern me much that we're seeing these types of fluctuations. And actually, being a geologist, looking at these invasions don't upset me too much, either. Lake Erie's only ten thousand years old, so every critter in it has invaded sometime in the last ten thousand years."

"Yeah," I said, "that's right. *All* species are invaders in a body like this."

"Yes. And they've all come to some measure of adjustment."

"So we're waiting for the zebra mussels to get adjusted," I said.

"They may have already," observed John Hageman. "In some cases."

"That'll ruin Melissa's research." I grinned at her. She laughed.

"It won't *ruin* it," she said. "These are *predictions*."

"There were a number of measurements of how many zebra mussels per square meter there were, in certain locations," John went on patiently. "And there would be some years that there'd be less than there were the previous year, and people would say, 'Well, what's going on here?' And one of the things they came up with was that, when you have extra high populations of zebra mussels, they possibly cannibalize their own veligers. So there may be a threshold of how much an area can support. When you have a dominant zebra mussel community of older ones, they can control recruitment of younger ones until they reach their old age, and die off, and leave room for new ones to start up again. They think they may be a little bit cyclic, too."

"Self-regulating," said Eddie. "It's gonna happen. At some point."

"People have gone to zebra mussel native range," said Hageman, "and have investigated parasites. One of the things they have found is a parasite that destroys the ovaries of the female mussel. And the populations in Europe may be in the thousands, but certainly not in the tens of thousands that we see here. They're really surprised to see 40,000 zebra mussels per square meter, when we have that all the time."

"And 350,000 in some places," I said. "Yeah."

John nodded. "Right—400,000 inside the Monroe, Michigan, power plant."

"You know, that environment seems real aberrant," Eddie put in. "I've never seen any other plant have a problem like that."

"All that material coming down the Detroit River from Lake St. Clair, apparently," John speculated. "You know, the Monroe plant sucks in 70 to 80 percent of the Raisin River some of the time— especially low flow. It's a huge plant for its location." He turned back to me. "We had a professor from Poland here—we showed her a typical rock with zebra mussels on it, and she thought we were playing a joke on her, because it was so covered. Her comment was that their average concentrations are only about 4,000 per square meter. And here we had 4,000 on a little rock this big." He held up his hands, thumbs and forefingers indicating an imaginary golf ball. "So we hope that they'll go back to their native range levels, of closer to 4,000 per square meter. And it may take sort of an inner mechanism for that to happen."

"I've done some diving on water intakes," persisted Eddie, "particularly the one in Huron, and we did some videotapes. It's a circular pipe about three feet in diameter, and there's a ring of zebra mussels around the edge—maybe three or four inches down. And then they've fallen off into the bottom, and so the bottom half of the pipe is sedimented over with fallen stuff. They live three or four years, and they die, and the byssal hairs eventually atrophy and they spald off. That can cause some problems inside the plant. But it seems, theoretically, that you're not gonna get much more than a ring of about four inches, and it's gonna be self-spalding, as the interior ones dry off. I don't know what happened to Monroe—it's the only one I've heard of where the plant got constricted to the point that there was a major water intake problem. That seems—"

"I was talking about the electrical plant," said John. "You're talking about drinking water."

"Well, whatever that plant was."

"It was the drinking water intake. It was a combination of mussels and frazil ice. You know, in the winter—"

"Could be, yeah," Eddie nodded. "I just remember that this one was a—you know, it caused—just—"

"Closed about three days," Hageman supplied laconically.

"But that's the horror story that you hear about. The rest of the intakes—"

"The whole community had no water for several days," said John.

"You don't hear the industry screaming much about it."

"Yes, chemical controls have been effective."

"Speaking of chemical controls," I said, "let's talk briefly about some of the *older* nonindigenous species, like the lamprey and the alewife. You never hear about them anymore. Everybody's talking about the zebra mussel. Have they gone away?"

"No," said Hageman.

"The lamprey has not," said Melissa. "We just heard a talk earlier this summer on the problems in the St. Marys River."

"And aren't they a problem in Lake Erie for the first time?" asked Eddie. "They've reached a threshold of concern."

"Well, they're certainly seeing more scarring on fish," said John Hageman. "The Lake Erie eastern tributary streams have been cleaned up, and farm runoff has diminished to the point where lamprey can spawn in some of the rockier higher-gradient streams."

"So cleaning up these streams has actually increased lamprey habitat," I said.

"Right," said John. "Most of these streams that have a viable lamprey population get a treatment at least every four years. Because that's the cycle you need to follow to effectively kill the larval lamprey."

"Why isn't it working in the St. Marys?"

"Too big. It's too big of a river. It's the major connection, as you know, between Superior and Huron, and it's just too much water."

"Well, how about the alewife?"

"The predators are keeping 'em under control," Hageman observed.

"How about the real old ones—the smelt, for instance?" I asked. "How are smelt populations?"

"Never were big here," said Eddie. "The big fisheries for them were over by Port Dover, on the eastern basin side, out of Canada. I've talked to people in Port Dover recently, and I don't think it's as big a fishery anymore."

"I used to come up here in the late seventies and early eighties," said Hageman, looking a little wistful. "And I can remember one year I took a garbage pail and dipped it into the water, and I had the whole thing full of smelt. Never seen that again. In the fishery literature, I think you'll find that Lake Erie smelt have some sort of a parasitic affliction that weakens 'em to the point where most of 'em don't survive their first spawn."

"So there's some hope for the zebra mussels," I observed.

"Yeah," said John. "And smelt's not necessarily a native to Lake Erie, either."

"No," I agreed, "smelt's an import. And carp was also a release."

"Yeah," said Eddie. "In 1858."

"*Way* back," I said.

"*Way* back," Hageman agreed. "Carp were good."

"We just got the wrong strain, I think," laughed Eddie. "Well, the cormorants—no, the cormorants are native, I suppose."

"Yeah," said Hageman morosely. "We were all celebrating five or six years ago, when the first cormorant nested in Ohio in a hundred years, on West Sister Island. Now there're thousands of 'em, and people are saying, 'How are we gonna get rid of these things?'"

"What was this shotgun killing the other day?" asked Dave Moore.

"Someone shot some?" asked Eddie.

"Several hundred were shotgunned to death," said Dave, "over in Lake Ontario. They don't know who did it."

"Probably a frustrated fisherman," said Eddie.

"The Lake Erie sports fishermen have a vendetta against 'em right now," said John Hageman. "They've hired a couple of our sea grant agents to go out and do some underwater video, to compare the fish communities in areas with cormorants and areas without, so you can see the difference. Which is a huge undertaking, with a video camera that basically runs in a straight line, and doesn't look both ways—you know—you're looking at a *little, tiny* piece of Lake Erie. But they said they'd try to do it. They got a free video camera out of the deal."

"Why would this be?" I asked. "Why would the cormorants be expanding their populations at the expense of other things now, when they've lived in harmony for—"

"Zebra mussels!" said John brightly. There was much merriment. "They're very efficient underwater, with the clear water now."

"That's right," grinned Dave Moore.

"It's come back," said Eddie, "by virtue of—well, that's the question we keep asking. What are the causes for this effect? What has caused the cormorants to come back? What has changed that now is conducive to these birds that wasn't twenty years ago?"

Melody had been looking thoughtful. "When you mentioned cormorants," she said, "I thought of passenger pigeons and buffalo. Some people now theorize they were in an explosive population phase that was *bound* to crash—it wasn't just Americans."

Eddie nodded. "Well, when you read Audubon's accounts of the passenger pigeon in the Cincinnati area there, in the Ohio River valley—I've always thought something was *wrong*."

"Yes," Melody agreed.

"I mean, there were too damn many." He turned to John Crites. "I felt the same way about mayflies, John, I thought that those populations were much too big for a stable situation."

I explained DuWayne Gebken's theory that good management techniques applied at the wrong time during a population cycle could actually end up causing more harm than good. Eddie looked interested. "What management strategy were they using?" he asked.

"He didn't specify."

"I was wondering, because there's really not a whole lot we can do. We can control phosphorus and nutrient levels, and we've done that a little bit—and we can control bag limit—and that's it. There's no other buttons I can push. Fish management, when you think about it—there's not a whole lot—"

"People management, is all it is," observed John Hageman. "Except it does include stocking, and control of undesirable fish."

"But stocking—you know, these hatcheries that we have here have shown that to be not particularly effective. On a large body of water, stocking doesn't seem to do much good."

Hageman nodded. "That's why they've given up on salmon, and are down strictly to trout. They expect their return at a specified time of year, and steer anglers toward those streams that they stock 'em in."

"These hatcheries here were set up to do whitefish, first, I think, John?" Eddie looked at John Crites, who nodded. "And then maybe to walleye—I remember when I first started here, in the sixties, there were walleye—"

"Yes," said Crites dreamily, "I remember walleye."

"Blue pike, sauger, and yellow perch," said John Hageman. "According to the signs inside the entry."

"I didn't know they ever did blue pike," said John Crites.

"Neither did I." Eddie looked doubtful.

"I read a journal that said they did blue pike," Hageman insisted.

Eddie looked at me. "Blue pike is our extinct species here," he explained, "that we feel sad about. There are people doing research now, retroactively, on the DNA. The latest we heard is that perhaps it was a true species of its own, rather than just a subspecies of walleye. There was quite a bit of genetic difference."

"People are wondering if there's blue pike in some places up in Ontario," said Hageman. "Lake Nipissing is one that's mentioned. They're doing DNA work right now to see if it matches the old blue pike that are in the specimen collections of the museums. And then there's debate whether—if they are—whether we'd want to stock them into Lake Erie again."

"Oh, I grew up fishing on them," protested Eddie.

"I'm all for it," agreed Hageman, "but the DNR's saying, 'Why wreck a good thing?' We've got walleyes that are providing a major fishery. Why do we want to give away some of our bait fish to blue pike that only get thirteen or fourteen inches—"

"Blue pike were more of a Central Basin fish, though," said Eddie, "which would enhance that part of the Lake, I think. They were a night fish. You'd fish for them at night, with a light. And it was a kind of a neat thing."

"Some people think that the ones that they're seeing now are just a color phase of the walleyes," said Hageman. "They may have some old blood—some hybridizations—but time will tell if they're different or the same."

"Well, refugia can develop in odd places," I said, "and you don't know where they are. And sometimes species hang on and then reappear."

"You know," Eddie mused, "I've seen a few more unionid shells along the shore, now, than I did the last couple of years. Just along the beaches where we live, and things like that. Not a lot, but enough to make you think they're still around."

"One of the things I heard," I said, "was that they are extinct in St. Clair. I don't know if that impacts things down here."

"They say they're functionally extinct here, really," said Hageman. "Might be a surprise refugium somewhere here that someone stumbles across, but if you go out to any of the clam beds that people would normally collect from you can search for hours and never find a live one. I'll give you an example. Off Kelleys Island—about 1989, or maybe 1990, when zebra mussels were first seen in Lake Erie—researchers wanted to go get some clams and do some controlled experiments, whereby they would put zebra mussels artificially onto a clam for half the individuals, and then leave the other ones naked of zebra mussels. The clams that were used for that experiment were collected in half an hour by one diver off Kelleys island. Two years later they tried to go back to the same clam bed, and the diver who had originally collected those eight hundred clams in a half hour took extra people with him, because he knew he was gonna have trouble finding them. And they searched for two hours underwater—about four or five divers underwater at the same time. And they never found a single live clam to bring up. So that population went from extremely abundant, to virtually, or possibly completely, nonexistent."

"At the very beginning of my parasitology classes up here," said John Crites, "I used to take people in rowboats and go right out the cable line. And we would free dive, and come up with six species of unionid mussels. We'd have more than enough to completely do anything I wanted to do with the class—if we needed more, we'd go back and get 'em." He shook his head. "That population's completely gone."

"And they'd grow sometimes along the shore, in the trough between the second and third sandbar," said Eddie. "You could stand in waist-deep water, with a large pail, and not move your feet, and completely fill that pail. Just from the ring you could make—"

"I did that," said John Hageman, "and made a batch of clam chowder, one time—back when they probably had more mercury than I needed to eat that night." We all laughed. John kept a straight face. "Well," he said, "I can tell the temperature, now. But I had this brainstorm. I'd walk out about waist-deep and feel with my feet and kind of lift them out of the sand, and just reach down, and grab 'em, put 'em in my pail—and that whole clam bed is now gone. Square miles of clams are gone. The last ones that we saw were on a small sand ridge at the entrance to Manila Bay on North Bass Island, and now even that's disappeared. But silver trout, which were on the endangered list, are now very common, to the point where a fisherman can catch several while they're out perch fishing. And about eighty sturgeon were caught last year in Lake Erie that were documented, either by commercial fishermen or sport anglers. So there's some fish that probably are taking advantage of—Oh! Zebra mussels!"

"We haven't touched on the spiny water flea," I said.

We don't see a whole lot of 'em in the Western Basin, where the fish population is the highest, so I guess it was self-controlling," said

Hageman. "People further east in Lake Erie complained that they foul up their fishing lures, and fishing lines, and downrigger cables, and such. But it's not an earth-shattering type of a problem. They're still around, and the fish can still find them. But we don't see the applesauce-like concentrations on the surface of the Lake like we did ten years ago."

"Do you think they've slipped into the ecosystem," I asked, "into a niche that's going to be self-perpetuating?"

"I can't answer that," said Hageman, "but they've found a better place in the Lake than the western end."

I asked how the group felt about the goal of zero discharge of toxic contaminants. "It certainly sounds wonderful, politically, for the EPA to put that out as a goal," Eddie said. "As far as being obtainable—I don't know. You go as best as your technology can do, and as long as an industry is using the best available technology, what more can you expect? Until the technology catches up."

"There are certain things that don't belong in the water, at any level." said Hageman morosely. "I think those types of operations should move elsewhere. They can send 'em out to Nevada or somewhere."

"Yeah," said Eddie, "but I think the thing is that, unless industry is banished—they can have the cleanest operation possible if they reclaim, and if that material has some value, why throw it out the door? There's a beryllium plant near here. And Ohio, for some strange reason, has dropped any beryllium standards, so that plant is dumping a lot of beryllium."

"There's slime from that plant that's turned the Portage River solid blue," said Hageman, "All the plants are blue—all the shoreline rocks are blue—"

"And the point Eddie made just a minute ago is very valid here," I said. "It's really not in the industry's best interests to be releasing that material, because it's *worth* something."

"Maybe what they're making is worth more," said Hageman.

"There have been win-win situations where they discovered they could fund their cleanup operations with the sales of the side products," Melody pointed out. "People have been encouraged to try to make more profits by finding things like that. But it's more cooperative, so it doesn't—"

"Well, yeah," said Eddie. "They gotta pay those lawyers—I mean, those lawyers they're paying have gotta do something."

John Hageman had been playing with his cap, turning it over in his hands. Suddenly he flipped it into the air, where it looped end-over-end twice and landed, perfectly positioned, on his head. Everyone applauded.

"I think John is telling us it's time to go home," I observed.

When we turned off the light it was dark—as dark as only a cottage in the woods, on an island, in the night, in the rain, can be. It was just two days from full moon, but not a sliver of moonlight penetrated the clouds, and there were no city lights, either. The rain pounded the roof, and the surf worried the shore, and the smells coming in the open window were cool and wet. There was a comforter on the bed, which we needed. Who needs air-conditioning when you have Lake Erie?

We almost missed Old Woman Creek.
That would have disappointed Eddie and Ricki, who were quite
adamant that we should see it. Old Woman Creek falls into Lake Erie
two miles east of Huron, where the Herdendorfs make their home
when they are not on South Bass. It is barely ten miles long and drains
a watershed of just twenty square miles—hardly imposing figures.
But there is one feature that sets this little stream strongly apart.

Ricki had suggested that we probably wouldn't spot the creek
until we were crossing it, and we didn't. One moment, U.S. 6 was
taking us through the residential edge of Huron; the next, we were
out on a bridge with the creek ponding against its barrier beach a
hundred yards to the north and a broad sheet of shallow, lotus-strewn
water, Old Woman Estuary, spread like a shout of nature to the south.
In perhaps three seconds we were across the bridge and into the
houses again, with only the fact that we had known beforehand that
it was there remaining to let us know that we hadn't witnessed a
hallucination from an earlier age.

We doubled back and found, just west of the bridge, the tiny
parking lot serving the rivermouth unit of the Old Woman Creek
National Estuarine Research Reserve. A dormitory for students and
resident researchers occupied a flag lot along the creek, behind a
private home: a path led down to the Lake. The sky was still gray
this early morning after our South Bass Island storm, and foot-high
breakers kept up a ragged drumroll against the lakeward side of
the barrier beach, which stretched all the way across the mouth
of the creek. It is this beach that makes Old Woman so special. In
presettlement times, in the dry season, barrier beaches were common:
it was possible to walk dry-shod across most of the river mouths
along the Ohio shore. The desire for commerce and pleasure-boat
anchorages has taken care of that. All the beaches are regularly cut

through these days to keep the harbor channels clear. All, that is, but one: Old Woman Creek.

Inside the barrier the water was calm and silvered by the reflected sky. Immense round leaves, covered with drops of water that danced like mercury, bobbed boatlike at the edge of the sand. A school of small silvery fish roiled to the surface, then disappeared when a gull screamed. At the far end of the barrier beach, at the base of a low bluff, a large sign proclaimed the land beyond it private; atop the bluff, half hidden by trees, was a house. It is odd what perspective will do to you. This time it seemed that the creek and the beach were real, and it was the house that must be the illusion.

We got back in the car and drove a mile east to the reserve's headquarters and visitor center, housed in a small wooden building on the shore of the estuary. Although mixing is somewhat inhibited by the sandy barrier at its foot, this is a true estuary, a drowned river mouth like that of the Mink back on the Door Peninsula. Here,

The bar at the mouth of Old Woman Creek, Ohio

however, the dominant wetland vegetation is not a sedge but the American lotus, *Nelumbo lutea*. Acres of lotuses lift their ball-shaped yellow flowers on three- foot stalks over big round leaves at the water's surface—the source of the discs we had seen floating near the beach. The effect is exotic, as though the viewer had suddenly been transported to India. We walked the Edward Walper Trail, a boardwalk that skirted the edge of the lotus beds and then climbed into blufftop oak-hickory forest. Red cardinal flowers brightened the understory; bright orange shelf fungi, looking a bit like large gumdrops, extruded from tree trunks. This is the land we have lost by becoming civilized. If you want to see what Ohio looked like when LaSalle sailed the *Griffon* along these shores, look here.

East of Old Woman the lakeshore was mostly houses. Tiny bedroom towns followed each other in bewildering succession: Oberlin Beach, Ruggles Beach, Heidelberg Beach, Mitiwanga, Volunteer Bay. Lake access points were few and far between. At Lorain we gave up trying and hopped on Interstate 90. Soon we were in downtown Cleveland. Traffic kamikazied past through a forest of billboards and buildings made genteely dirty by decades of steel mill emissions. We took the Carnegie Avenue exit and rode the tidal wave of traffic east along Carnegie and Cedar to Cleveland Heights. A big, familiar brick church loomed on the right. We turned left, onto a quiet street of large trees, small green lawns, and tidy, well-kept houses. In a very few minutes we were sitting in Ed and Anna Fritz's backyard, in lawn chairs, in the shade. Ed pressed a bottle of beer into my hand; Anna brought out the cheese and cold cuts. The sun dropped under the clouds and beamed brightly on what was suddenly a much calmer, quieter, and altogether homier Universe.

A confession: I would have insisted on stopping to see Ed Fritz even if he hadn't been one of the people I had interviewed about the Great Lakes in 1983. You might not have read about it here, but we would have made the stop anyway. An elongated German leprechaun, tall, skinny, and redheaded, Ed has a habit—disconcerting to some—of following ideas through to their logical conclusions, which may or may not be where the originator of the idea thought it would end up. He is a chemical engineer by vocation. Anna is a social worker. She is short, neat, and businesslike, with a softly clipped accent that reveals its origins in the South Africa of her youth.

And then there was Mosey, who was a surprise. Mosey was large and long-haired and strikingly patterned in black and white, and he was purring his way into Melody's lap almost before she could sit down. I hadn't thought of the Fritzes as animal people, and said so. Ed smiled. "Our friends tell us we're typical late-life parents," he said. "Mosey has pretty much taken over the place." We drank some more beer. Suddenly it was past midnight. We went to bed in a small room up under the roof, turned a box fan toward the bed, and slept as if the concept of sleep had just been invented.

THE TEMPERATURE HAD CLIMBED INTO triple digits by late the next morning, and it stayed up there till sundown. With Ed at the wheel of Anna's Honda Civic we headed for the mouth of the Cuyahoga River. The idea was to start at Lake Erie and go upstream as far as we could comfortably get in an afternoon. I did not expect the bottom end to be pretty. Ed and Anna assured me that, further up, it would be much better.

Like Old Woman Creek, the Cuyahoga has a symbolic weight out of proportion to its size. Barely one hundred miles long, with a drainage basin of just over eight hundred square miles, it not only isn't the largest Lake Erie tributary, it doesn't even crack the top five. Among water quality activists, however, the Cuyahoga has a reputation more famous—or infamous—than any of its larger brethren. It is the River That Caught Fire.

There were actually not one but several fires, going back at least to 1936. All of them took place in the area known as The Flats, a heavily industrialized floodplain stretching from the river's mouth on Lake Erie roughly eleven miles upstream to the Route 17 bridge. Most were fairly small: it was the big one on June 22, 1969, that caught people's attention. Flames as high as five stories shot upward from the Cuyahoga's surface that day, destroying one railway trestle and heavily damaging another, and the fireboat *Anthony J. Celebrezze* had to be brought upstream from the Cleveland waterfront to put out the river. *Time* magazine ran a feature on the flammable river, and those attempting to clean up Lake Erie took the story and ran with it. It didn't seem to matter that the fire actually burned oil and grease and floating surface debris, not river water, or that the principal problem with the Lake was nutrient enrichment and algal blooms, not industrial waste. Rivers are not supposed to burn. When they do, it is clear evidence that something has gone horribly wrong.

Northern Ohio residents, who know and love the Cuyahoga in

all its moods, have long since tired of jokes about asbestos canoes and pre-cooked fish. "That was a non-event! It didn't happen!" Ed sputtered back in 1983 when I tried to show Larry Chitwood, in Ed's company, where the fire had taken place. It was not the historical account of the fire he was disputing—that would have been difficult to deny—but its significance to the state of the river. "What matters most," he emphasized that day, "is not the things that float on the surface. It's the things dissolved in the water column." Those, and the ones in the bottom sediments, do not burn. By focusing on the weenie roast on the water's surface there was danger that we would overlook the dark beasts slithering about underneath. The point is well taken, but it ignores the symbolic weight of flames rising from the seemingly unburnable. It was a movement, as much as a river, that caught fire on the Cuyahoga. There had been efforts made to clean up Lake Erie before. It was the one that flowed from the burning river that succeeded.

We began our tour at the river's mouth, at the bottom end of the restored and yuppified Flats, reclaimed from their industrial-wasteland status and now supporting restaurants and fashionable nightspots. I had already seen some of the changes here. In 1983, Ed and Larry and I had stopped for a beer in a little tavern called Fagan's, which featured two rooms, an old battered bar, and a huge fireplace. A new deck overlooking the river was evidence that the water quality was improving, but the day we were there the deck was closed due to heavy, gritty dust blowing in from a nearby pile of taconite (pelletized iron ore). In 1990, when Melody and I and our daughter Sara passed through on Sara's way to Hampshire College in Massachusetts, Ed had said with a twinkle, "I have to show you something," and then he and Anna had taken us all to lunch at Fagan's. In seven years the place had been utterly transformed. A vast acreage of dining tables spread behind a wall of fashionably slanted glass facing the river. Ferns hung from the ceiling. The food and the menu—and the prices—were elegant.

The taconite pile was still there.

In 1998 there was no need to stop at Fagan's: been there, done that. We passed quickly through the yuppified lower portion of The Flats and nosed on upriver into the still heavily industrialized upper end. Little macadam roads, curbless, curved every which way. Ed threaded his way through them unerringly. Steel mills and castings plants and rusty railroad trestles loomed over us; the river oozed past between black cindery banks, and there wasn't a sprout of green to soften them. Melody compared the scene to J. R. R. Tolkien's Land of Mordor, and she didn't get any argument from the two Cleveland residents in the car, either.

South of Rockside Road the change was abrupt. The narrow, swift river was lined with trees and grasses; it curled through fields and forests and wrapped about the bases of tall clay bluffs. The water was pale green with reflections of foliage. It was translucent, not transparent, and I certainly wouldn't drink the stuff, but it would have been comfortably canoeable. Within a few miles we entered the Cuyahoga Valley National Recreation Area (NRA). Tree-shaded picnic areas sprawled beside the river; bicycle trails lined its banks. Despite the heat, they were full of cyclists.

The NRA's principal bicycle path, the Towpath Trail, threads the park from end to end along the route of the towpath for the old Ohio & Erie Canal. The ruins of the canal itself run like a parallel riverbed beside the Cuyahoga, sometimes watered, sometimes dry, almost always visible. The 308-mile-long Ohio & Erie operated from 1832 to 1913, carrying both passenger and freight barges between Lake Erie and the Ohio River. Forty-four locks lifted it from Lake level at Cleveland to the Portage Summit in Akron, nearly four hundred feet higher. Mules hitched in teams of three towed the barges, weighing as much as eighty tons each, along the flat stretches between the locks. After a flood in 1913 halted operation of the canal, the towpath fell into disrepair. The twenty-two miles of it through the National Recreation Area were revived as a bike path in 1993; supporters envision it as part of a future trail stretching the entire 308 miles from downtown Cleveland to the Ohio River.

We stopped at Lock 29 and walked south for a short distance. The towpath was thronged with other walkers and with bicyclists. Most of the park's use comes from the Cleveland/Akron corridor, but we encountered one young woman from California, and there were license plates from New York and Wisconsin in the parking lot. Deer damage was evident in the woods along the trail. Like many other parts of the country, the Cuyahoga Valley is fighting a losing battle with Bambi. In the absence of predators—including human hunters, who are banned from the park—the deer population has exploded. Park authorities attempting to control them have felt the wrath of animal-rights activists, who have so far managed to block all efforts to reduce the size of the herd the only way it can be reduced—by killing some of the deer. "I tell people it's a choice between letting the Park Service manage the park and letting the deer manage the park," Ed said. "So far the deer are winning."

We stopped to examine Lock 29 itself. The Ohio & Erie's locks were built of locally quarried sandstone blocks, carefully shaped and set and often marked proudly with the mason's initials, and most of them held up well during the sixty-year period of neglect between the canal's closure and the NRA's formation. "It's just marvelous construction," said Ed, stroking the knife-thin joint between two of the blocks. "I often wonder how they were able to do it with the primitive tools of the day."

"They had the basics," Melody pointed out. "The lever, and the inclined plane, and—"

I tried to extend the list. "The lever, the inclined plane, the screw— there are four basic machines. What's the fourth?"

"Irishmen," said Ed.

At a roadside produce stand in the upper end of the NRA (this is not your father's national park) we bought freshly harvested corn and home-baked strawberry shortcake. Back in Cleveland, Melody and I shucked the corn in the backyard, in the shade, while Mosey prowled around our feet and Anna put chicken on the grill. We drank some more beer. The warm darkness came down around the cooling house, lighted, even here in the city, by faint pinpricks of stars.

At breakfast the next morning, Ed looked aghast when I told him our proposed route, along Ohio 283 through Willoughby and Eastlake and Mentor-on-the-Lake. "You won't see anything but urban sprawl up that way," he informed us.

"Precisely," I said.

He grinned. "Oh, yeah. I forgot."

Anna was wearing a T-shirt that said "RESILIENT" across the front; I was wearing one that proclaimed "GROWING OLD ISN'T FOR SISSIES." We posed for pictures together. Melody and I said our good-byes— including a protracted one to Mosey—and headed for the urban sprawl.

First, though, there was the Chagrin River, deep in a green gorge marking the eastern end of Greater Cleveland. The morning air was clear and benevolently warm; a thick canopy of oaks and maples hung over the twisting two-lane road, backlit brightly by the sun. The river was lined with farms and parks. It was prettier than the Cuyahoga, although you shouldn't tell Ed and Anna I said that. Cleveland is lucky to have it, and those few Clevelanders who know about it are even luckier that most of the city's residents don't seem to be aware of it at all.

All too soon we were back among buildings. Malls sprawled; traffic oozed and sogged. Detours sent us down residential streets not meant to handle the numbers of vehicles the detours imposed on them. The nearby Lake didn't seem present even as an idea. I thought of Francis Parkman, writing in the introduction to the 1892 edition of *The Oregon Trail:* "The sons of civilization, drawn by the fascinations of a fresher and bolder life, thronged to the western wilds in multitudes which blighted the charm that had lured them."

Something of the same mechanism has gone on along the shores of the Great Lakes. Where is the Lake at Mentor-on-the-Lake? I saw strip malls. People are drawn to the coastline for a view of water, and a few lucky homeowners get one. The rest might as well be in Kansas. Houses jostle elbow to elbow along the water and spread back from it, covering all the available surface. Zebra mussels ain't got nothin' on us. The bucolic little coastal villages of eastern Ohio are covered with thousands of foreign shells, like the clams in Lake St. Clair, and like the clams they are dying under the extra weight. The fact that the shells have three-car garages and four bathrooms each adds to the scale of the problem but does not change its basic nature.

The situation is compounded by our ignorance of geologic forces. We tend to think of geography as immutable. Lakeshores are not. Sand washes away; the mouths of rivers move; bluffs crumble beneath the onslaught of waves. We should not build permanent structures on these transient landscapes. But over and over we do, and over and over the landscapes shift and the structures fail, and over and over we demand that governments stop it and engineers fix it. This is futile. Governments today are no more capable of commanding the sea to stop rising than they were in Canute's time. And there is nothing broken, unless you count our powers of observation and our common sense.

At the Grand River there was a break in the sprawl, and a road led north to Headlands Beach State Park and the Empire Dunes. Dunes have always been rare along Lake Erie, and these are just about the only ones left. It was lovely to see the tan swells of sand, crowned by marram grass, standing against blue water beneath the bluest sky we had seen since Tawas Point. The midmorning beach was not yet crowded. Little waves swept in and smacked on the pebbly sand.

We wandered down to the water, and for the first time I saw what all the fuss over Lake Erie's water quality was about. We had been observing Erie, you will remember, under gray skies and storm conditions, and the water had been as cloudy as I had remembered from previous visits. Today the weather was clear, and so was the Lake. Astoundingly, vibrantly clear. Sunlight polished the transparent water and illuminated the pebbles on the bottom. The rift flowing up the beach was visible only by its sparkles, and by the dark wash it gave to the sand. It was easy to understand Eddie Herdendorf's childish delight in the clarity of Put-in-Bay Harbor; easy, even, to forgive scientists who should have known better for referring to it as "oligotrophic."

But of course it isn't oligotrophic. The clear water here has not been caused ecologically, through the classic combination of deep water, cold temperatures, and nutrient-poor runoff: it has been caused mechanically, through filtration. The fact that the filtering has been done by a biological agent, the zebra mussel, does not really change things. The same could be claimed for filtration by any of the methods employed by that most ubiquitous of biological agents, the human being.

Now: having established that this biological scourge, the zebra mussel, has a beneficial side effect—and ignoring, for the moment, the probably accurate observation by Eddie Herdendorf and John Crites that the mussels shouldn't get all the credit—are there similar beneficial side effects one can point to from the similar biological scourge of suburban sprawl? Perhaps so. John Hageman had pointed to one possibility during that long conversation at Put-in-Bay: the tendency of newcomers to an area to be strongly protective of the amenities that brought them there. The Empire Dunes and the Cuyahoga Valley and Old Woman Creek are probably safer with houses crowded around them than without, because the people in the houses know them and love them for what they are rather than for their ability to generate money for someone. Does this benefit mitigate the harm the homes the newcomers live in cause? No more than the clarity of Lake Erie mitigates the harm caused by the zebra mussel. But recognizing it may at least allow us to keep our perspective.

We stopped once more in Ohio, at Ashtabula, another of the International Joint Commission's AOCs. This one takes in the lower two miles of the Ashtabula River, plus the harbor and part of the nearby lakefront. A tiny tributary called Fields Brook appears to be the primary source of the river and harbor contamination; most of it is included in the AOC, too. John Hartig had remarked that all the AOCs in Ohio were doing well, but they are not all doing *equally* well, and this one is probably the worst. There is a stakeholder's committee in place, and a RAP has been prepared and approved, but little work has been done. And the plan, as written, doesn't even deal with Fields Brook: that has been turned over to Superfund. As of the time we passed through Ashtabula, no remediation had been started. Cleanup of Fields Brook was expected to begin in 1999: the harbor and adjacent areas would not be attacked until 2000 or 2001.

I hadn't visited Ashtabula in 1983, so I had no previous impressions to go by. It turned out to be just as well, because we couldn't see the harbor anyway. The Highway 531 bridge over the Ashtabula River was out, preventing access to the roads that would have given us the views I wanted. We did drive down to Lake Shore Park, where turnouts along the bluff gave us partial views of the harbor facilities. Beyond piles of sand, which may have been harbor dredgings, we could see the tiny old Ashtabula Harbor lighthouse. That was as close as we got. Lake Shore Park has a boat ramp, and it was in heavy use the whole time we were there: boats were coming and going in the park's little basin at a fairly steady rate. Given the proximity of the AOC, I had no real desire to join them. I have no idea whether the ramp was open because the water was clean enough to allow it, or whether closing it was simply too hot, politically, to consider. The latter has certainly happened often enough to force consideration of it as a serious possibility. Even when there are good reasons for a rule, people tend to see its imposition as a violation of their rights rather than as a tool to protect them. Back at the Tip of Point Pelee, we had seen a big sign—present in 1983, as well—which commanded "NO WADING—DANGEROUS CURRENTS." Approximately every third visitor to the Tip wades anyway. We are a clever race, but we are not often very wise.

At Erie, Pennsylvania—the Keystone State's narrow window on the Great Lakes—we drove out the long, complex peninsula of Presque Isle Park. Beaches on the Lake side that I remembered as narrow, disappearing strands were broad and full of people. A stop at one showed the reason: a series of dolos, short seawalls with gaps in them, sitting in the surf several hundred feet offshore. They were working; the beaches showed that. They were also ugly. Was the cure worse than the disease? Only Erie's residents can judge. Guessing from the popularity of the beaches, I would say they didn't think so.

On into New York along Highway 5, the closest through road to the Lake. The sandy bluffs we had been seeing along Erie's coastline gave way to small shale and sandstone cliffs. Little streams stairstepped their way down to the Lake over low ledges in vertically walled gorges. They looked mythological. Where the sea cliffs began, sprawl largely stopped, as it had on Michigan's Thumb: coasts where no one can reach the water are not high on developers' lists. There were farms and wineries. The cliffs, where visible, were gorgeously grim. Beneath them the big blue Lake spread to the horizon. This is rural land, not wild, and aside from the Lake itself there is nothing in the scale of the landscape to justify the term "spectacular." What it is, instead, is warm and inviting. Fecund, even. If Cleveland's Flats are Mordor, this is the Shire. I hope the hobbits don't mind sharing it.

All too soon we were approaching Buffalo. Fields and vineyards gave way once more to housing developments; towns grown too big for their boundaries elbowed each other along the lakeshore. Traffic clotted in the streets. Highway 5 passed through Mt. Vernon and Athol Springs and Lackawanna and threw itself into the air as the Buffalo Skyway. The narrow, high-speed freeway looked dizzyingly down on decrepit factories and decaying harbor facilities. In *The Late, Great Lakes* I had compared the view from the Skyway to one of Hieronymous Bosch's visions of Hell; aside from a quick glimpse of the Tifft Nature Preserve just as we started up the Skyway's western ramp, there was nothing visible today to alter that opinion. I doubt God will send anyone from Buffalo to Hell when they die. Why bother?

At the eastern end of the Skyway we dropped like a stone into the heart of the city. Traffic flew by like errant bombs; signage was terrible. We blundered our way through, cursing at all wrong turns, and eventually managed to find Highway 266, the road down the east side of the Niagara River. The river flowed broadly and serenely on our left: no hint of the fate awaiting it a few miles north. The riverbank was lined with parks, and although they did not look terribly well kept we were happy to see them. It really *is* the thought that counts.

Past Tonawanda and North Tonawanda, almost into the city of Niagara Falls, we finally found the motel where I had arranged reservations. The building at the front of the grounds which looked like the office turned out to be a private residence; the building at the rear which looked like a private residence turned out to be the office. They couldn't find our reservations, but since it was approaching 8 P.M. and another party had not yet arrived, the manager decided we could have their room. I got a look at the paperwork and discovered, with no real surprise, that the tardy party whose room we were taking was us. I think that was the point at which I gave up trying to understand Buffalo.

Morning dawned thick and muggy and hot, with a faint smell of chemicals riding the northerly breeze. We went looking for breakfast and found waste dumps. First came the 102nd Street Landfill, the infamous little brother of Love Canal, fenced and capped; the vent pipes standing up from the grassy mound of the cap were an obvious source of the chemical odor we had detected earlier. Then came Love Canal itself. I recognized it easily, though it had changed dramatically since Jane Elder and I had toured it in 1987. Many of the houses had been reoccupied, and where these predominated the streets looked pretty much like any Sunday morning suburb: cars parked at the curb, sunlight filtering through maples and oaks, people out jogging, old men in bathrobes fetching the Sunday paper. Then you would turn a corner and there would be a whole row of empty houses, or an abandoned school, boarded up and forlorn. In the middle of things squatted the dump, mounded and capped like the one at 102nd Street and glaring balefully at passersby from behind its surrounding cyclone fence. The cap bristled with vent pipes; even with the car windows up and the air-conditioning on we could detect the sick-sweet odor of chemicals. This slow venting of gases is necessary: otherwise they would build up under the cap and eventually rupture it, with unforseeable but almost certainly dreadful results. So the dump's neighbors get chronic low-level exposure to chemicals rather than acute high-level exposure. This is an improvement, I think. I'm not sure anyone really knows. Despite years and years of education and legislation and technological innovation, our culture still does not handle its garbage very well.

We spent most of the afternoon at the Tifft Nature Preserve, the green space just south of the mouth of the Buffalo River we had glimpsed briefly from the Skyway the day before. In heat and humidity as high as the haze-filled sky we prowled marsh and woodland and felt renewed. Tifft is stark proof that we *can* do the right thing with garbage if we try: despite its name and its generally wild appearance, this nature preserve is a reclaimed dump. Once there was a farm here, but that ended well over a century ago. By the 1880s the farm had become an iron and coal transshipment facility, busy with docks and railroads. In the 1940s the city of Buffalo took it over and used it as a municipal landfill for a while, pouring garbage in and around the old transshipment harbor and several on-site ponds and wetlands; when these got too full, sometime in the late 1950s, the city followed what was at that time standard practice and simply abandoned the place. For fifteen years the space that had once been the Tifft farm lay idle. Vegetation came slowly back around the ponds, and birds and wildlife found it. So did birders. The word slowly spread.

Around 1972, with the environmental movement in full swing, Buffalo decided to do something about its abandoned landfill. The first plan was to blade it flat and use it for playfields, but when city officials announced this they rapidly found themselves under attack. That would destroy the ponds and the reemerging wetlands and woods, and there were far fewer ponds and wetlands and woods in the greater Buffalo area than there were playfields. Quickly shelving the playfield idea, the city left the garbage piles in place and simply sealed them beneath a five-foot-thick layer of impermeable clay, creating a series of natural-appearing mounds around and among the site's water features. The erstwhile farm, transshipping facility, and city dump reopened as a nature preserve in 1976.

The tree-lined parking lot for the preserve faces Lake Kirsty, a small, shallow body of water whose riprapped near shore is the only immediate clue to its artificial nature, although its equally spaced right-angled arms give away its origin as a harbor with quays extending into it. We found a parking spot in the shade and strolled into the visitor center, a rustic wooden building with the feel of a northwoods lodge. A covered porch faced the little lake, which was ringed by cattails and lively with waterfowl. Beyond, old roadways converted into paths led through woods to the edge of a seventy-five-acre marsh. A boardwalk, the Heritage Trail, looped into the marsh amid shoulder-high cattails. Songbirds made trills of themselves; turtles sunned on slanted sticks. A jumble of decrepit buildings and storage tanks looming to the north, in the Buffalo River AOC, seemed curiously disjointed from the view: Tifft felt far more concrete than the actual concrete of the decaying waterfront that surrounded it.

At the edge of the marsh we met Bill, whose last name I learned and then immediately forgot: I have thought of him ever since simply as Bill the Guard. A trim man of about fifty, with merry eyes and a small clipped moustache, he was sitting on a bench by the water, a bicycle leaning beside him, binoculars in his hand. The twin stripes down his dark shorts and the patch on the sleeve of his light blue, uniform-style shirt marked him as park staff; the square brown plastic badge pinned to his pocket said "Security." I asked him, after introducing myself and my mission, what he thought had changed at Tifft in the thirteen years he told us he had been there.

"Not much," he said. "Things have grown a little, that's about it. Of course, the muskrats are gone . . ."

Nobody knows where the muskrats went, but their population crashed about eight years ago, and cattails, once limited to the marsh margins because the muskrats ate them, have since spread over most of the marsh. Foxes have also declined. Water levels are up—the marsh connects to Lake Erie through a pipe four feet in diameter—and they've had to build a new, higher boardwalk (he pointed out the old boardwalk to us, mostly inundated, directly beneath the new one). And zebra mussels have invaded through the connection to the Lake. "People throw bottles in the water," he said. "You pull them out, and they're just covered with mussels."

Finally, Canada geese have become a serious problem. "I actually saw someone petting one the other day," he marveled, shaking his head in disbelief. "People feed them all the time. They stand right in front of the signs which say 'Do Not Feed the Wildlife' and toss bread to them. You tell them not to, and they're doing it again as soon as you turn your back." He laughed. "One lady was very irate when my administrator told her to stop. I thought it was going to lead to fisticuffs right there in front of the nature center. You know—'You can't tell me not to feed the birds. I have a right to feed the birds. They're hungry.' " Bill thought U.S. Fish and Wildlife Service policies might be causing some of the problems. "They're setting bag limits by the production of geese in Canada," he told us. "But a lot of geese don't migrate—they just move in year-round where they're being fed, or where nesting boxes have been set up for them. They aren't being counted in Canada, which drives the quotas down. But there's lots of birds. They've really become pests in many places."

An important lesson to draw from this is that major changes can take place without people noticing them, especially if they're incremental. The old frog-in-boiling-water technique clearly works on humans, too. Bill the Guard didn't think anything much had changed until he started detailing the changes—and even then, he may not have considered them major. The marsh was still a marsh, after all—sunshine, water, cattails, birds. What had really altered?

Fish leaped in the open-water part of the marsh as we talked, and small frogs plopped about over the damp bank. Sandpipers (another recent addition: there was no shorebird habitat before the water level rose and inundated some flat terrain, turning it into mudflats) squabbled over hunting territory. Bill, a birder by hobby, clearly loved what he was doing. He had once had a desk job, but the plant he worked at closed and he was laid off. Overall, he thought that was a good thing. His wages were smaller here, but he was much more satisfied.

"I'll leave you to your work," I said as the conversation drew to a close.

Bill leaned back on his bench in the sun. "My work is so hard," he smiled.

As we approached the car, we noticed a gull at the edge of Lake Kirsty which had evidently swallowed a fishhook: a juvenile, it was gagging and drinking a lot of water, a short length of fishline dangling from its beak. Melody pointed it out to another security guard, a young black man with a shaved head and a matter-of-fact but pleasant demeanor, and after studying the bird through binoculars he went and got a net to catch it and haul it to a veterinarian. While he was gone, it moved across the pond to the top of the pipe leading under the Skyway to Lake Erie.

"Oh, man," said the guard when I pointed out the bird's new position to him.

A bystander asked if he was going to catch it. "Not over there, no," he replied. "Over here I might have a chance, but not over there. It'll just fly." Bill the Guard rode up on his bicycle about then, and the younger guard went to confer with him. We got in the car and left.

Tifft was a happy exception. But I'm afraid that most of the time we were in Buffalo I felt, like that juvenile gull, as though I was gagging on a fishhook. I apologize to Buffalo residents who love their city.

214

There are probably good reasons to love it, but to this outsider it was not a pleasant spot. We blundered our way through the city once more, pretty much giving up on signs after the fourth or fifth misdirection. The street map we had purchased the evening before was not a great deal more help. Rand McNally is known to place small inaccuracies in their maps as a means of detecting illegal copies, but the copy-protection department appeared to have gone overboard on this Buffalo map; it seemed riddled with errors. (Just now, as I was writing this, Melody came into my office looking for a map of Portland, Oregon. I had the Buffalo map in my hand at the time, and I offered it to her as a substitute. She snorted. "That would probably be about as much use in Portland as it was in Buffalo," she said.)

Eventually we found the motel again. It was only a little past four in the afternoon, but we had no taste for further exploration. We were carrying a lot of reading matter with us in the car, and we are still pretty good company for each other after thirty years of marriage, and the room had a decent television set. And that's how we spent one summer Sunday in 1998 in Buffalo, New York.

On Monday morning we had to go into the city again. "All ashore that's going ashore," I said, squaring my shoulders behind the wheel.

"Batten down the hatches," said Melody.

"Man battle stations."

Clearly we did not want to go back into Buffalo.

But the trip went smoothly this time—probably due, in no small part, to the fact that we pretty much ignored the street signs—and we had a couple of good conversations. The first was at the offices of Great Lakes United in Cassety Hall, on the campus of Buffalo State College. I asked the organization's biodiversity field coordinator, a redheaded young man named Andy Frank, how GLU was faring.

"Pretty stable, right now," he told us. "It had been through quite a turbulent period right before my involvement—they had gone through several executive directors in just a few years. It's hard to maintain stability that way.

"It's a unique place, in its mission as a coalition. There's always some energy going in different directions, and it's hard to pull it all together at times. There's a lot of interest at the public level, and interest by the funders, in issues of sprawl, and biodiversity, and habitat fragmentation, and habitat loss. At the same time, the commitments made to toxics and cleanup under the Water Quality Agreement haven't been fulfilled. So as a group of activists working on the Basin—how much do you let go of something that hasn't been finished to pursue something where there's a lot of energy? There's always dialogue going on. I would say there's a good chunk of it going on around that issue right now."

The GLU coalition includes both environmental and labor groups, which often have trouble working together. I asked Andy if this split personality was causing them any problems. "Not on very many issues," he said. "Our traditional constituency of UAW and

Canadian Auto Workers doesn't have a lot of conflict with most of the types of things that we address. And there's room for growth, I think. Part of the work with the unions has been on a concept called 'Just Transition'—directly addressing the issues of what's going to happen to workers in a move toward cleaner production, as opposed to saying, 'Well, deal with it.' " I think development of that concept is another thing that lends credibility to the environmental movement. Not just being blindly opposed to everything."

"Are you working on sustainability issues at all?" I asked him. "Does 'Just Transitions' tie into the question of sustainable workforces and resources?"

"I would say yes, we're 'working on sustainability,' " he smiled, "but not in a fully strategic enough sense yet. There's a binational toxics strategy that was signed a little over a year ago by the Canadian and U.S. governments, and we're involved in that, helping to develop some projects in the Great Lakes region that eliminate rather than reduce some of the persistent toxics like mercury and dioxin. There's some changes in language that a lot of people in the environmental community are uncomfortable with, by the way. It's not necessarily 'zero discharge' now, it's 'virtual elimination,' and how 'virtual' is defined is a little vague. And this new toxics strategy has very little teeth, so it's really a discussion around sustainability in its most fundamental sense—around what industry says is achievable, and what other folks say is achievable. You have to engage those dialogues at some point, and hash that out. And on forestry issues there are a lot of dialogues happening around sustainability. Around two issues, basically—having adequate forest interior habitat, and also having a sustainable agricultural wood supply. To oversimplify, I think pulp demand is the problem. Hardwood timber can be managed reasonably in a lot of ways, but pulp demand is heavy, and it's hard to have our consumption of pulp still on the rise while being able to preserve enough forest to support caribou and moose and wolf. Canada has enough now, but the way their government is moving—and as the demand for pulp, and for just wood in general, is rising—you know, that's *fleeting*."

"I see a tie-in," I observed, "between the question of pulp—if that's the problem—and the toxics issue. Because a lot of the toxics come out of the paper mills."

"Right," agreed Andy. "It's a definite link. It's separate in one sense, in that there are chlorine-free processes, so the industry that produces paper and wood products could be a lot less polluting. But there's a link. Societally, you have to address paper consumption, and recycling, and alternative fiber crops—there's a lot of things to tie into in that whole discussion. And I think that all the components of that are being discussed, but they haven't been brought together yet, in terms of getting the public really to understand all those connections, or in getting the people in decision-making power really to understand."

I asked Andy, who grew up in Buffalo, what I was beginning to think of as the Change Question. What, in his opinion, was the most important change of the last fifteen years?

He thought for a moment. "Well, the most important thing was getting control of the nutrient pollution," he said, after a bit. "That was one of the obviously detrimental things. Getting control of that has been a good thing."

I smiled. "But that's partly been done by the zebra mussel."

"Well, the extent that it's as clean as it is has been done by the zebra mussel—OK. We're gonna go there too." He laughed. "But actually, they've hit their targets in terms of what they were trying to do with phosphorous load reductions. We still have problems with combined sewage overflows and direct discharge—some beach closures—Woodlawn Beach, in Buffalo, closes regularly because of that. But there was a basic achievement in terms of visible water quality. And there's been some reduction in the discharge of toxics, too, but not as much." He sighed. "Toxics is so fuzzy, because industries start introducing different chemicals that aren't necessarily reported as they change their processes around. And people don't see toxics as much—they're pretty much invisible. I think there's a public perception that things are much cleaner than they are. The fish advisories are still strong—the concerns about mercury in fish are alarming. There was some cleaning that was achieved, but I think there's a perception that it's much stronger than it really is.

"But beyond that, what has happened over the last fifteen years, in terms of the ecosystem of the lakes, has been exotics. As you alluded, the zebra mussels. The alewives had moved in before. The alewife populations were *so* high, and there was not a predator on them, so they would eat themselves to a point where they just crashed and died, more or less, and there were dead fish all over the beach. So there've been all these transitions that have gone on in terms of introducing non-native sport fish, Pacific salmon, to feed on the alewife—and now, fifteen years later, you have a group of users who want to maintain a Pacific salmon fishery in the Great Lakes, that can't support it naturally. They need the alewives for the salmon to feed on. So now you're managing a non-native prey fish for a non-native sport fish."

"Are they actually managing alewives these days?" I asked. "I've heard rumors to that effect."

"Yes," Andy said. "They're not stocking alewife, or anything like that, but they are tracking the population and trying to understand its fluxes, to make sure that they can support a viable Pacific salmon fishery. Especially in Lake Ontario—their management plan is designed around that. But I would say it's an aspect of all the Lakes.

"So you have intentional non-natives that are affecting the Lakes' ecosystem in a dramatic way. You also have unintended exotics. Zebra and quagga mussels—there's other preyfish, the ruffe and the round goby, which have started to impact food chains themselves. There's just so many different things caused by that, in terms of impacts to the ecosystem. There's a lot of benthic crustaceans that are starting to be displaced. It's not definitive that the zebra mussel is causing that, but we certainly know that they're taking up the available plankton faster than anything else. The available productivity of the Lakes is sucked up primarily by the zebra and quagga mussels, and not effectively recombined into the ecosystem. They're mostly not eaten by other fish."

"Does that have an effect on the planktivorous fish as well?"

"Yeah," he said. "In Lake Erie there's a call right now by angler groups to cut back on sewage treatment of phosphorus, because the populations of perch and walleye are in decline. It could be because there's not enough prey—could be because the water's too clear, and it affects their hunting—there's different theories. And the angler groups are saying, 'Well, you know, we think the solution is to just add more phosphorus.'"

"They're actually saying that." I needed to hear it confirmed.

Andy nodded. "Not that that won't just lead to more zebra and quagga mussels—but that's their notion of a solution. So it's gotten a lot of press. The Ontario Federation of Anglers and Hunters is one of the most vocal advocates, so it's been a very big Canadian story. It's incredible to folks that have a history of the Lakes, who've been trying so hard to get phosphorus reduced for so long."

"That's reminiscent of a few years ago, in the late eighties," I mused. "All the cottager groups were complaining that the Lakes were too high—the Corps of Engineers had to lower them immediately. And then natural processes came along, and they lowered two feet."

"And it was too low," smiled Andy.

I nodded. "And they started screaming that the Corps of Engineers had to *raise* the Lakes—"

"Right," he said. "We're having a little bit of another wave of that, because the levels are high again. I just got a notice for a meeting on the Lake Ontario shore about water levels. So there's a lot of dialogue starting to happen, and that cycle of overreaction, and still not understanding how the process works, is happening all over again."

"That ties into suburban sprawl, so let's talk briefly about that," I suggested. "That's an issue people have brought up regularly. Do you follow it at all?"

"Yeah," he said. "My purview is habitat and biodiversity. I focus on the fisheries ecosystem, forests, and wetlands—that's an oversimplification, but that kind of approach. And especially in wetlands work, urban sprawl is *the* issue. To a certain extent in forest issues as well, sprawl is *an* issue. But I would say that river and stream and wetlands habitats are all primarily threatened by sprawl. That is their major threat right now. Again, there's kind of an illusion that things are under control with wetlands permitting. People know that there's a permitting process. But"—he looked thoughtful—"there's a project that we've been working on in Utica—it's basically a mega-Wal-Mart development. They call it a 'super power center,' but what it is, is a Wal-Mart with several other major retail outlets attached to it."

"Strip mall," I observed.

"Yeah, a mega–strip mall. A strip mall not of little shops but of megastores set next to each other. The developer in this case believes that, because this project is a 'super power center,' by definition it needs a massive footprint. And this massive footprint has to sit on top of the stream and all of the wetlands at this site. They can't scale down the project—if they did that it wouldn't be a 'super power center,' and they would have no project. That is the argument they're making, essentially." He shook his head. "Not 'essentially'—it *is* the argument they're making. They have a stream that spawns trout, and they believe they have a case for putting buildings and a parking lot right on top of it. They think—'Well, we'll mitigate. We'll rebuild the stream over here, and we'll build some new wetlands over there.' They believe that that's adequate, that that's OK, that that'll take care of things. And a number of politicians believe that's OK, too. In this case, there's been involvement of the local congressman, and some of the local leaders, as well, who can't understand why the Corps is being so—difficult about this plan. The Corps is saying, 'No way. You can't have the stream and the riparian wetlands.' They just don't understand it. They've been calling the federal EPA and the federal Corps offices, and saying, 'This Buffalo district of the Corps is just unreasonable.'" He grinned. "So there is a big problem. Humbug Marsh—I don't know what the most recent information is that you have, but that area is protected under a conservation easement that was signed as a mitigation for a project that somebody *else* did several years ago. And now they're coming back and saying, 'Well, OK, so we signed this as a conservation easement for the rest of eternity for the general public, now it's five years later, how about we do away with this conservation easement so we can build a development in this last marsh of the Detroit River.' You know, that it's even being *considered* tells you how weak our regime is. The mantra is 'No Net Loss.' There's this assumption among politicians and among developers that mitigation is a panacea, and it can work in all cases—it hasn't been demonstrated to work very well in even the *simplest* of wetlands. There's nothing you can do about bogs and fens. How are you going to replace a mature forested wetland?" He sighed. "So there's this idea that our regime will be around 'No Net Loss,' and we can use mitigation to get to 'No Net Loss.' And if we believe that, we're just going to lose systems—more and more systems."

"We're talking a bit about the role of government agencies here," I observed. "How do you see their role having changed in the last fifteen years?"

"Well, I think the argument to downsize and reduce regulatory burden has been successful, in a significant way," Andy stated. "So that even the more mainstream, more environmentally supportive politicians have a lot invested in the notion that there must be simpler ways to do things. And at the same time there's been a lot of disinvestment in enforcement. It's gone through cycles in New York. When Governor Pataki was first elected, the folks that they had in the DEC [Department of Environmental Conservation] were very corrupt, and I think under Engler's administration as well, in Michigan, there was a lot of major corruption of the regulatory process, of making these decisions political rather than based on what the law required, and what was known scientifically about what the resource needed. There's been a lot more of that. It seems to have overstepped its bounds, some, and it's getting reined in a little, but I wouldn't say that we're anywhere near moving forward, in the Basin, with really progressive protection regimes."

"What would you say are the changes that will be seen in the *next* fifteen years?" Melody asked.

Andy looked thoughtful. "I don't know—that's a tough call. I think that the notion of sustainability will have to get addressed in the next fifteen years, on a lot of different levels. Or else we'll have gone way further than we can really afford. Our appetite consumes all over, and our style of living is spreading to a lot of other nations. Everybody's consuming, and we haven't figured out better ways to do stuff. So I think that set of issues will need to get addressed, just by virtue of what has become obvious is excedence of what the ecosystem can bear. I guess that's the way I'd frame it."

We left the Great Lakes United offices and drove under darkening skies to the sprawling suburban campus of Buffalo State College's big brother, the State University of New York at Buffalo, where I had scheduled a conversation with Joe DiPinto. Joe is the director of SUNY-Buffalo's Great Lakes Program. The GLP operates out of Jarvis Hall, a big concrete box that also holds the university's engineering department. In what seemed typically perverse Buffalo style, there was no building directory to be found. Staircases ran up the outside of the building: they appeared to ascend all the way to the roof without encountering any doors except the ones at the bottom. But there *were* elevators. We picked one of the elevators at random, took it to the second floor, and by dint of blind luck—certainly not by any help from the signage—walked right into the GLP office.

Joe kept us cooling our heels for fifteen minutes while he spoke with some of his students. He apologized for his tardiness.

"That's all right," I told him. "The students come first."

He smiled. "I'm going on sabbatical at the end of this week," he said. "Everybody seems to need a piece of me before I go. Come on in."

We went into his inner office, which was in the disarray that comes with hurried packing. Joe settled down behind his desk. An intense, solidly built man with a bald head and a dark beard, he bore a striking resemblance to the man who has been called "the voice of the Great Lakes," Canada's beloved, prematurely deceased folksinger Stan Rogers.

I asked him the Change Question. He thought for a moment.

"I guess what I'd like to initially talk about is the quality of the Lakes in terms of toxic chemicals," he said after a bit. "In general, the concentrations of what we call 'bioaccumulative chemicals of concern' in the Lakes, and in the biota in the Lakes, are decreasing dramatically. Twenty years ago there were probably eight to ten parts per million PCBs in lake trout in Lake Ontario. Today—same age fish, same Lake, et cetera—the PCB levels are probably more like about *one* part per million. One to two, maybe. I think that comes from a combination of things, if you want to think about what's caused it. One, we obviously have put a lot of efforts toward eliminating the use of these things. We've sunsetted the manufacture and use of PCBs, at least in this country. We dug up a lot of capacitors, and things like that, and removed them as sources of PCBs. That has affected it considerably. And the other thing that I think has really helped is the significant effort we put into reducing phosphorus loadings to the Lakes. That had a side benefit of also reducing toxics, because when you're removing particulate matter more efficiently in your municipal wastewater plant, you're also removing a lot of the hydrophobic chemicals, like PCBs, that are associated with particulate matter. So the discharges from municipal treatment plants to the Lakes, of not only phosphorus—which was the main goal—but of toxic chemicals, have decreased considerably."

"Nothing to sorb to," I said. "Yeah."

"So that's the good news. Right now, though, with regard to toxics, we're at a point in the Lakes where the decrease has leveled off considerably. It's still going down, but it's going down much slower than it was through the eighties."

"I've heard several theories on why."

"I think it's a combination of things," said Joe. "I think the Lakes are close to being in steady state with the existing loads. We've decreased the loads considerably, and the last 10 percent of what we originally had is the hardest, obviously, to decrease. A lot of it is atmospheric in nature, and a lot of that atmospheric input is actually chemicals from outside the Great Lakes Basin. So that's part of it. I think also part of it is that the years and years of abuse, of heavy inputs of these chemicals, have stored a great deal of them in the bottom sediments of the Lakes. We're getting to the point now where the natural resuspension and redeposition of sediments that occurs because of wind-driven resuspension creates a buffering system that prevents the concentrations in the water column from decreasing as fast as they might otherwise."

"Are we talking about deep-bottom sediments here?" I asked. "Because the harbors have been mostly cleaned."

"We're talking about—yeah, depositional areas, largely," he clarified.

"Outside the harbors."

"Right. In the Lakes. But not all of the harbors have been cleaned."

"Well, of course," I said.

"Speaking of harbors, another thing that's occurred in the last fifteen years in the Lakes, that was not around when you wrote your book, is RAPs. Remedial Action Plans for Areas of Concern."

"The AOCs were in existence," I said, "but nobody was paying any attention to them."

"Well, they hadn't really started doing anything with them," he said. "And we've done a lot since, but—you know, if you actually look at them, of the forty-three Areas of Concern, only one has been delisted. So we have a long way to go. Cleaning up historical in-place pollutants, in sediments, is very, very expensive. And we're not putting in nearly enough resources, in my opinion, to do that job.

"A lot of what's happened, though—over this period of time— is that we have removed the sources. And what we're seeing is a natural cleanup in the harbor areas. We've shut off the inputs. We're still getting particulate matter coming down from the watersheds and depositing in the harbor areas, and the lower portions of the rivers, but those solids are now *clean*. So we're *burying* the contamination. While we're sitting here talking about it, and trying to figure out where we're going to get the money to do the cleanup, it's been occurring naturally." He laughed. "I guess if we wait long enough we won't need any remediation. But—I don't *know*. Anyway, RAPs are an interesting phenomenon."

"The RAP process has done *this*," I said, describing an arc in the air. "It's gone up, and then it's come back down."

"It has," Joe agreed. "That's to a large extent because the governments are putting less resources into it."

"For whatever reason."

"It makes it very difficult for the general public to maintain momentum without resources. I mean—you can only beat your head up against the wall for so long." He laughed again. "Anyway, the other thing that's happened is that fifteen years ago we had not achieved the phosphorus target loads quite yet. I would say maybe thirteen years ago we got there, so we were close. But the phosphorus targets have now been achieved. And what I think is really worth mentioning is that those target loads, which were set in the late seventies—twenty years ago—were set on the basis of mathematical models. Those models were used to say, 'If you want to achieve *this* level of chlorophyll *a* in the Lakes, then you need to reduce the total phosphorus load in the Lake to *this* amount.' They used the models to set the target loads, and then spent a lot of money upgrading particular plants, and by the mid-eighties we'd achieved them. And people went out and said, 'OK, now you've got to the target loadings—did it really convert into the chlorophyll *a* concentrations? Are they at the levels that we thought should happen, on the basis of the models?' This is what we call a 'post-audit process.' You predict something with a model—you predict what a given remediation will do, and how the system will respond—and then you implement that remediation. Then you go back and see how well you did in predicting it. And those models were almost right on for these systems. It was really, I think, a very important success story that I hope you can tell. Because it did look good."

"In fact," I observed dryly, "I just heard that it looks so good that some of the angler groups are lobbying—"

"Well, I was about to say—*until.*" He grinned. "Until zebra mussels came along and threw all those relationships out the window. That's another area of *major* ecological changes in the Lakes within the last fifteen years—actually, within the last *ten* years. And that, of course, has caused significant changes, especially in the Lake St. Clair and Lake Erie ecosystems, because with the introduction of the quagga along with the zebra mussels, it's almost completely covered the bottom. I mean, *really* high densities."

"One of Ed Herdendorf's graduate students has found them on soft bottoms as well as hard bottoms," I said.

"Right," said Joe. "And the quaggas seem to be better at colonizing soft bottoms, and in deeper waters." He leaned forward. "What zebra mussels do is very interesting. They increase the daily flux of particles out of the water column. They're tremendous, very prolific filter feeders. As a result, they have diverted a lot of carbon and energy from these systems into the bottom sediments. The water clarity of the Lake has increased tremendously. Secchi depth readings in the Western Basin have gone from one or two meters up to six, seven—"

"I've heard—yeah, seven or eight," I said.

"It's almost as if all the people in Lake Erie can see is that they *can* see," said Melody.

Joe laughed. "And, you know, you would think that that was good," he said. "But in the late eighties, while we were spending tremendous amounts of money trying to figure out how to keep zebra mussels from growing in water intake pipes, we weren't spending enough money trying to figure out what sort of ecological implications they were having in the Lakes. Subsequently we figured out how to keep them from growing in water intake pipes, and sort of solved that problem. But it's only in the last five years that we've really started paying more attention to the ecological end."

"I noticed an article on the bulletin board in your outer office that said property values had gone *down* because of the higher clarity of the water," I said. "That amused me."

"Well, it's because of the macrophytes," said Melody.

"You get a lot more macrophyte growth," Joe agreed. "If you go down here to the Erie Basin Marina here in downtown Buffalo, there are macrophytes growing in there now that never were a problem."

"I understand the mechanism," I said. "It's just—you'd think—"

"Right," he said. "Right. And there's some other nonintuitive impacts that are occurring as a result of zebra mussels. In the Western Basin of Lake Erie and in Saginaw Bay we're seeing late-summer blue-green algal blooms. Which we thought we got rid of, with phosphorus control. Which we *did* get rid of with phosphorus control, and they're coming back. With no change in the phosphorus loadings in the Lakes. I've been doing some modeling work in Saginaw Bay to try to understand what's going on, and it seems, based on our work, that it's a combination of things. Zebra mussels selectively reject blue-green algae. They spit 'em back out. They have the capability to do that. I don't know if it's a taste response, or what. But based on our modeling work, that is not enough, in itself, to give us late-summer blue-green algal blooms. In addition to that, what it takes is a shot of phosphorus right at that time of year. What we think is happening is that by pulling the diatoms and the green algae out of the water column in spring and early summer at a very high rate, and repackaging them as feces and pseudofeces, they've altered the phosphorus cycle in the system. The phosphorus that comes in the spring, with the spring runoff, gets taken up by the algae, and then they get pulled out of the water column into the sediments. And this stuff sort of hangs around, and then in the late summer, in August, when things warm up—when conditions, in terms of temperature, are good for blue-green algal blooms—you get this mineralization of phosphorus back into the water column. Just at the right time. And there's no competition, and now they get this shot of phosphorus— and that's causing blooms. They have a competitive advantage, in that they're not grazed at as high a rate, but they also are getting this phosphorus that wouldn't have been there if it hadn't been for the zebra mussels."

"Is it coming at a slower rate but still coming from the streams, or is it coming from release from the feces and pseudofeces through some other mechanism?" asked Melody.

"I think it's actually a release," Joe said. "It's a mineralization from the feces and pseudofeces, which are now part of the bottom sediments. Bacteria work on it, or something, and in the late summer, when the temperatures are warmer, this happens faster. It's kind of an interesting phenomenon. People blame a lot of things on zebra mussels, but this is one that's real." He grinned. "And the other thing I've found that's really interesting is that you would expect—if you start with the fact that zebra mussels pull particles out of the water column into the sediments at a higher rate, you would say, 'Well, maybe that's good, in terms of pulling toxic chemicals like PCBs out of the water.'"

"But they don't seem to do that, do they?" I observed.

"They don't," he agreed. "It's interesting. I've done some screening-level modeling in Lake Erie that suggests that they do decrease the total PCB concentrations in the water column somewhat. But not as much as you would think. The total does go down, and the concentration of PCBs in the bottom sediments does go up, because of zebra mussels, so the benthic food chain is more susceptible to bioaccumulation. And that's sort of an intuitive thing that you would expect to happen. What's *not* intuitive is that the *pelagic* food chain— the open-water food chain—is *also* susceptible to higher bioaccumulation rates. Why? Because—we've seen this, and we predicted it in the models—when you've got less particulate matter in the water column, you have a much higher fraction of the total chemical in the dissolved phase. Which is the bioavailable phase.

"Say you've got a total PCB level in the water column of ten nanograms per liter. If you've got ten milligrams per liter of suspended solids, then a very high fraction of those PCBs, maybe 90 percent, are going to be on the particles. And the concentration per mass of particles will be low, because there's lots of particles around. But when you've only got *one* milligram per liter of solids for that same ten nanograms per liter of PCBs, you're going to have a higher fraction in the dissolved phase, and the concentration per unit mass of particles is going to be much higher. Our calculations suggest that if you continue to load Lake Erie with the same PCB loading, but you add zebra mussels—versus not adding them—you can as much as *triple* the concentration of PCBs per unit mass of algae in the water column. And of course algae is the base of the food chain. We haven't actually done the calculations of bioaccumulation, but that suggests that there's a lot more of it. Assuming that the predators in the water column—you know, the upper trophic levels—continue to feed at the same rate."

"And it is happening, isn't it?" I observed. "The greater bioaccumulation?"

"Well," Joe said cautiously, "you're seeing PCB levels in walleye in Lake Erie—it's hard to say, we haven't got enough years to say there's a real trend, but it looks like it's at least *not dropping.* It may be going up. Historically, walleye in Lake Erie derived most of their food from pelagic fish, like smelt. It's possible that walleye may now be getting a higher percentage of their energy from the benthic food web, rather than pelagic, just because there isn't as much. Which leads to the next big issue, I think, in the Great Lakes, and that's fisheries relative to water quality. I think we're seeing some tradeoffs now between fisheries management and water quality management. Historically—I mean probably fifteen to twenty years ago—those two groups weren't talking to each other very much. I think one of the big things that's occurred in the last fifteen years, in addition to RAPs, has been the realization—we're still not there in terms of actual implementation, but the realization, anyway—that we need to take an ecosystem approach toward managing the Great Lakes."

"Officially, that came in in '78," I observed.

"Yeah," he said, "but I think it's only in the last five or ten years where we've actually started to do something that's reasonable. We're starting to realize that management of one particular issue or problem in the Great Lakes does have implications on another. In Lake Ontario, for example, the fishery is based on stocking. We're getting some lake trout back, but mostly we're stocking salmonids. I think Lake Ontario's a prime example of how, up until the last ten years, those two camps were not talking to each other. Because at the same time we were reducing phosphorus loadings to Lake Ontario, and the total phosphorus concentration in the Lake was dropping, we were increasing the stocking of salmonids—lake trout, coho, chinook, brown trout—by a factor of two or more. A tremendous increase. Because what they were seeing was—they stocked more—fishing got better." He laughed at the tautology. "But now what's happened is, we're at the point where what we call the 'bottom-up effect'—low productivity at the base of the food chain, resulting in less food for the forage fish—has combined with so much predation pressure caused by stocking so many top predators that we've put the alewife, which is in the middle, at tremendous risk."

"There was a time a few years ago," I observed, "when that would have sounded very odd."

"That's right," he said. "That's right. Because we were trying to get rid of 'em. They were washing up on beaches. But now, as it's turning out in Lake Ontario, they are basically the only forage fish in the Lake. Which is not a good thing. We need more diversity. But, anyway, what finally happened in '93 was that they decided to reduce the stocking by a factor of two, roughly, to reduce the predation pressure on alewife. They recognized that there is a limited carrying capacity of the Lake for fish, based on the amount of nutrients that are falling into it. And they've continued that, although there's pressure by the charter boat industry to push it back up—they don't think it's as bad as other people think."

"Of course," I said.

"And the same thing's happening in Lake Erie, with walleye. The walleye fishing community is screaming that we need to—titrate phosphorus back into the Lake—" He grinned. "I think the scientists' general feeling is that that would not be a good idea, for a number of reasons. One is that it could exacerbate this blue-green bloom problem with the zebra mussels. Another is that we're not convinced that walleye are actually *limited* by nutrient loadings. It may be habitat."

"So that gets into the issue of urban sprawl," I said.

"Yeah," said Joe. "Wetlands—spawning areas—may be the real limitation for the walleye population. But the other thing that's happened with walleye is that the people on Lake Erie have been sort of spoiled. Ten or fifteen years ago the walleye population was really high. It's come back down in the last five or six years. But really, when you look at it in the long term, it's come back down to levels where it was in the seventies. It's now back maybe to normal. And that's causing problems, and cries for more phosphorus loadings."

"We're assuming that phosphorus is the limiting nutrient," I pointed out.

"It is," Joe said. "For algal growth. For primary production. It *is* the limiting nutrient. It is not always the limiting *factor*—light and temperature also play a role. But nitrogen, actually, has gone up considerably over the last fifteen years."

"Years ago I heard that in Green Bay nitrogen might be the limiting factor."

"There were times in Lake Erie when nitrogen was limiting. But today we've engineered these systems to phosphorus limitation. And you rarely, if ever, see blooms of nitrogen-fixing blue-green algae. The blue-green blooms that we're seeing now, that we think are zebra mussel ecology, are non-nitrogen-fixing blue-greens. You don't see the kinds we used to see back in the seventies. That's a real good indicator of nitrogen limitation, because those guys get a very obvious advantage under those conditions."

"Do you have any theories on the decline of yellow perch in Lake Michigan?" I asked.

"I was on a review panel for Wisconsin sea grant proposals last fall," he said reflectively, "and there were several proposals to look at that. It may be ruffe. I've heard that ruffe is sort of taking over the niche. I think that's a possibility, and I think also zebra mussels—" He laughed at his own suggestion.

"Everybody blames zebra mussels for everything," I observed.

"Sort of like El Niño, you know?" he grinned. "You can blame everything in the weather on El Niño, and everything in the Lakes on the zebra mussel. But it is possible that they're diverting energy from the perch."

"One question I've had," said Melody, "is what happens with—well, I wrote it down as 'zebra biomass decay.' You answered a lot of what happens with their excreta, but I don't know what happens with their bodies."

"What happens when they die?" asked Joe.

"Yeah," said Melody.

"You see empty shells," I pointed out. "You don't see what emptied them."

"Yeah," he said. "Well—a couple of things. I was just talking to somebody at Stone Lab, and they've found a lot of zebra mussels in gobies. The gobies apparently are doing a good job. So there are some predators."

"It takes an exotic to catch an exotic," I remarked. "Is this the idea?"

He laughed. "I guess. We have seen the zebra mussel densities sort of come down and level off since their first introduction. And that's typical of organisms. When you bring 'em in they really overshoot their carrying capacity, and initially they go up very rapidly, and then they come back down, and—"

"Some species have their own population moderations," Melody observed, "and some don't."

"Right. Uh—and I think—zebra mussels don't." We all laughed.

"But we're starting to see them come back down. I think its a combination of the fact that they originally overshot considerably, and also we're starting to see some predators find them. I think it's starting to develop. But there has been some concern that zebra mussel decay is what's causing the taste and odor problems."

"We've been blaming it on algae," I said.

"Well, historically, yeah. And I still think that's where the problem is." He laughed again. "But I've heard some people blame it on zebra mussels, too. It probably could happen if the mussels were actually growing in intake pipes."

"Where are you planning to go on sabbatical?"

"I'm going to Michigan. I'm staying in the Great Lakes. I'm going to be working at a consulting company in Ann Arbor, Michigan, called LimnoTech, that has the same goals I have for aquatic ecosystem modeling. Like I said previously, historically the people that were looking at eutrophication ignored the top of the food chain, and the people interested in fish ignored the lower part of the food chain. And the people that did toxics ignored *all* of it." He grinned. "I think the next generation of modeling in the Great Lakes will couple all these various factors in a single framework, so that you can actually operate the kinds of feedbacks that we've been talking about earlier, where, for example, zebra mussels can generate blue-green algal blooms. So my sabbatical is to sort of continue that development, so that we can really hope to understand the indications and responses of these systems—the multiple responses to multiple stressors acting in concert. What I'm doing on sabbatical is trying to develop an expertise in aquatic ecosystem modeling."

When we came out of Jarvis Hall it was raining. We had come to the western end of Lake Erie in the rain: we would leave the eastern end the same way. We blundered our way northeast from the university to New York 78 and headed north out of Greater Buffalo without even an inkling of a regretful look back.

At Lockport, Route 78 ran briefly beside the Erie Canal, still in operation—though much altered—170 years after its construction. Then it turned north and dropped over the Niagara Escarpment. Ten miles to the west, the waters of the Great Lakes were making the same descent, surrounded by tourists and shills: having done that scene several times before, we elected not to bother. Niagara Falls is a spectacular, massively powerful, incredibly awesome display, and everybody in America should see it. But I wish they wouldn't all try to see it at the same time.

At the foot of the escarpment the highway arrowed out onto the coastal plain of Lake Ontario. Farms and orchards slid past; damp little villages loomed up and fell away in the rain. Just south of the tiny hamlet of Newfane a fine and fair-sized river with the undistinguished name of Eighteenmile Creek came swinging in from the southwest and bent to join our course. A few miles further north, at Olcott Beach—which the map labeled "Olcott" ("Rand McNally is copy-protecting its maps all over the place," remarked Melody)— Route 78 bumped up against the water and stopped. A town park stepped down the low bluff to the beach. Children were swimming, and several adults sat on folding chairs watching them, in swimsuits, in the rain.

We backtracked two blocks to Highway 18 and turned east, along the lakeshore. Farms and orchards pressed close against the damp two-lane blacktop: farm stands offered fresh peaches. Old farmhouses, some looking as though they dated from before the Revolution, lined the road. At Golden Hill State Park we stopped briefly and dashed down to the cobbled beach. Fireweed brightened the bluff below the old Thirty Mile Point Light, built in 1875, out of commission since 1958; gulls swam, rather than flew, overhead. The Lake faded into the rain a few hundred feet offshore.

Here I had better make another confession: I find Lake Ontario considerably less Great than its four brothers. Volumetrically it is four times the size of Lake Erie, but in area it is less than three-fourths as large, and somewhere in that drop from 9,900 square miles to 7,300 a threshold seems to be crossed. We are a predominantly visual species, after all: we take most of our cues for the feel of things from the look of them. And Ontario just looks smaller. Some of the clues are subtle: less surf, and fewer signs of its effects along the shore. (Wave size and power depend on *fetch*—the distance the wave has traveled across the surface—rather than depth or volume.) Other signs are more obvious. When Larry Chitwood and I camped here at Golden Hill in 1983 it was a clear evening, and as the dark descended we could see a light twinkling far out across the surface of the Lake. Consulting a map revealed the light's source: the aircraft-warning beacon on the CN Tower in downtown Toronto, on the Canadian side. The CN Tower is more than a third of a mile high, and that, of course, has something to do with it. But when you can look across a lake and see a building—even a very tall building—on the far shore, something about it shrinks. Lake Ontario is a very lovely body of water. It is even a very *large* lovely body of water. It is the fourteenth-largest body of fresh water in the world. But it is not in the same league as its four big brothers up at the top of the hill.

Smaller size does have its compensations, though. One was apparent as we drifted east from Golden Hill along gray roads, as close as we could stay to the shore. Though this stretch of Great Lakes coastline is within an hour's drive of nearly a million people in Buffalo and Rochester and their circles of suburbs, it is only lightly developed. There were the usual lakeshore cottages, side by side, but here they were commonly old and small. Sometimes they were new and small. Sometimes the "cottage" was just a slide-in pickup camper without its pickup, propped in the middle of a green lawn ending at the lakeshore bluff. It was only as we approached Rochester that the massive new developments that had been so common elsewhere around the Lakes began cropping up again. Lacking a better explanation, I am going to have to assume that it is the smaller, more intimate feel of Lake Ontario that has produced the smaller, more intimate scale of development here. It is probably significant that by the time the big developments pick up again the Lake has bulged slightly, from forty miles wide to about fifty-five. There is far less chance, there, to get an errant glimpse of Canada across the water.

We hit Rochester at rush hour, in heavy rain, and it was still better than Buffalo. Slipping out of the flow, we followed residential streets, staying as close as possible to the water. The mouth of the Genesee River sported a small park with a fishing walk. Further east, the city's large green Durand-Eastman Park was soggy and dark, though it was just past 5 P.M. in the middle of the summer. Stopping the car, we made a dash for a large tree which promised to ward off the rain. From its shelter we could look down a low bluff to a beach and a small sliver of Lake that disappeared quickly into fog. A woman with a raised umbrella strolled beside breakers that were rapidly getting damper. I recalled the people sitting in beach chairs in the rain back at Olcott Beach a few hours before. It seems to take a great deal to deter New Yorkers who really want to get their money's worth out of Lake Ontario.

Our Rochester waterfront odyssey ended at Seabreeze Park, in an old, upscale suburb on the western point of land enclosing the mouth of Irondequoit Bay. We took the Seabreeze Expressway south along the bay and merged into the eastward-flowing rush of Highway 104. Soon we were in Webster, where we had motel reservations. The motel was new, and it was sparkling clean. The window of our second-floor room opened into a copse of twilit beeches in which a female cardinal was playing about. The rain had stopped, and the clouds had begun to lift. We were not out from under The Line yet, but we were clearly getting very close to its edge.

WE AWOKE EARLY TO THE SOUND OF birds: the night in Webster had been cool enough to turn off the air conditioner and leave the window open. The sun was shining, and the sky had a scrubbed-clean look. I pointed the car toward Lake Ontario along fresh and sparkling residential streets. Quite by chance we came to the coast at a large county park where lush grass swept down a bluff beneath great old trees to the edge of the chopping water. We walked out the park's long breakwater. Waves pounded the west side; to the east, in the lee, the Lake's surface danced lightly, bobbing a small flock of mallards gently up and down like rubber duckies. A golden mist hung along the shore, barely heavy enough to be seen, like a veil left over from some bright enchantment at the dawn of time.

East along the edge of the Lake there was a great deal of development, but there were farms and orchards, too. We followed Monroe County's Lake Road across the line into Wayne County, where it became County Road 101. Private communities appeared, gated and full of large, close-set houses. They bore names like Lakewood and Edgewater, and most of them were new since 1983. The tiny hamlet of Pultneyville, however—whose Pig Inn Road gets my vote for the best street name in America—looked very much as it had fifteen years ago. The overall effect was mildly schizophrenic: big new developments, tiny old cottages, and behind them all the great blue sheet of the Lake, both older and newer than anything along its shore.

County Road 101 ended at Sodus Point, the complex three-pronged sandbar that guards the western approach to Sodus Bay. The point was covered with grand old frame cottages fronting small, winding streets—very old-money, very upscale, and with lots and lots of "Keep Off" signs on the beaches. That's a characteristic of our way of doing things, and it always seems wrong. Why should access to an international treasure like the Great Lakes be bought and sold to private individuals? We have these massive, biologically crucial, scenically stupendous gems of fresh water, and we insist on carrying them to market like a carton of eggs or a box of paper clips. And we don't even have the good sense to be ashamed of ourselves. Not enough of us do, anyway. I enjoyed our rainy night on the shore of South Bass Island with Eddie and Ricki Herdendorf, certainly, and I wouldn't want to take Castle Park away from Amy Blossom and her family. But a line must be drawn somewhere, and when it is, I think we will find that we crossed it long ago. The Fallacy of Composition is as inexorable as it is inescapable.

Sodus Bay, however, was a delight; and Little Sodus Bay, up the road another twenty miles, was even more so. These eastern Lake Ontario embayments are drowned stream mouths, surrounded by wooded hills and carrying little islets: they appear to be excellent sailing waters, if the number of small-boat sails we saw was any indication. White frame houses decorated the shores like accent marks. It is undoubtedly hypocritical of me to complain about private ownership of the shores in one breath and praise the appearance this private ownership has created in the next, but what the hell? The sun was shining, the sky and the water were blue, and the curve of shore and splash of houses and white scatter of sails formed a picture that was very close to perfection. Emerson was right about foolish consistencies and the hobgoblins of small minds.

The core of Oswego did not appear to have changed, but the outskirts had picked up a crush of big-box stores and a flutter of fast-food outlets that left no doubt about the passage of fifteen years. I looked for, but failed to find, the road to the Black Lagoon, a Superfund site I had visited in 1983: things in the east part of town, where the site lay, were simply too different to be easily recognizable. The site had been cleaned and capped, anyway, and the EPA had crossed it off its list, so there would quite probably have been nothing much to see there even if I had found it. We really are getting a handle on toxics, at least from the point-source and small-area source perspectives.

It was past noon now, and the air was beginning to feel oppressive. We followed state highways around the southeast corner of the Lake to Selkirk Shores State Park. The surf was up: trains of two- to three-foot waves were rolling into the park's swimming area from the limitless blue horizon. It was almost as though Lake Ontario was daring me to keep thinking of it as less than its larger brothers. Swimmers, mostly young teens, were leaping the waves and bodysurfing on them, and it was a tossup whether the surf or the laughter of the surfers was the louder sound.

A short distance north of Selkirk Shores the Lake Ontario dune belt began. There was not much of it—less than twenty linear miles—but while it lasted it was pretty much a match for Lake Michigan's. Dunes and barrier beaches enclosed ancient bays, converting them to ponds and wetlands. Small streams ran, rush-lined, toward the freshwater sea. At least that's what we saw in passing. In common with most of the lowest Great Lake's shoreline, there were few points of public access. We drove into Sandy Pond, the small dunes community where the EPA's Bob Beltran grew up. It turned out to be a cluster of cottages sitting on the biggest set of dunes on the Lake and not letting anyone from outside near. There is a Nature Conservancy preserve at the tip of the Sandy Pond spit, but we saw no access to it: I have been told, since, that access is only by boat or by walking two miles up

Swimmers bodysurfing in Lake Ontario at Selkirk Shores State Park, New York

the beach. These shores should be the birthright of all of us, not just those who have real estate deeds in their pockets. The New York state parks, for the most part, don't help much: they contain marinas and swimming beaches and big grassy lawns to camp on, and that's about it. Selkirk Shores had a picnic area called "The Bluff." It was tree-lined and cyclone-fenced, and there was no view of either the Lake or the bluff the area was named for. Is this what New Yorkers really want?

Lake Ontario's dunes are not only locked off from most public access, they are disappearing. The river of sand that feeds them has dropped to a trickle, and no one seems to know precisely why. In fact, although research is under way as I write this, know one even seems to know precisely where the sand comes from. All that is known for certain is that it is no longer arriving, at least in sufficient quantities to feed the dunes. Much of the seventeen miles of sand beach here has turned to cobble in the last few years. Dunes are a dynamic environment, and this may be a natural phenomenon; but in this culture of groins and riprapped shorelines and tightly pinned down coastal real estate it seems likely that human activities will turn out to be the principal cause. We have placed most of the shoreline in a cage. We should not wonder that those few bits we have left outside are having trouble reproducing.

Thunderheads were gathering in the west as we left Sandy Pond. We had planned to camp somewhere in the Thousand Islands, but a look at the sky made a motel seem prudent. We hopped on the freeway and hurried north. The room we found at Alexandria Bay had a slight scent of mildew, but everything worked, and here in the heart of one of New York's great tourist regions at the peak of the season it was a steal for sixty dollars a night. The back window offered a view of a clear pond lapping against glacier-polished stone. At one end of the pond a great blue heron statued through a small wetland. I was beginning to like northern New York motels. We locked the door and went out to look at the St. Lawrence River and the islands they named the salad dressing after.

Native Americans called the Thousand Islands the "Garden of the Great Spirit." It is easy to see why they were so enamored of it. This is land ground down to its elemental stone heart and then inundated. Small polished islets rise out of clear water: woodlands decorate them, looking extruded rather than grown, as if rock and trees had been born together out of the depths of the St. Lawrence. On the larger islands stand houses, rarely more than one per bit of water-surrounded stone. Such is the magic of this place that the houses seem natural, as fit to their setting as the stone itself.

The storm that had been threatening since Sandy Pond broke over the islands as we lingered at Keewaydin State Park, just outside Alexandria Bay. Black clouds boiled downstream past the International Bridge, bristling with lighting. Thunder hammered the water. It was like being inside an immense Van de Graaff generator. In the campground, campers were placing lawn chairs for the best views. "This is better than TV," one shouted to us as what looked like a wall of electricity crackled and slammed over the granite-girdled river. The first drops of rain fell as we stood on a concrete fisherman's walk by the water, watching boats bound toward cover. Big drops, like bursting water balloons. We dashed for the car as the heavens opened. Curtains of falling water obscured our view of the campground, where the campers were no doubt rapidly putting their lawn chairs away again. I wondered if the guy who had made the comment about the TV had a set installed in his shower.

Through what seemed like a car wash gone berserk, with sound effects, we drove upriver toward the tip of Cape Vincent, where the old Tibbets Point light marks the beginning of the St. Lawrence. Beyond the town of Cape Vincent the road narrowed. Heavy development— all expensive, mostly new—lined the left side of the narrow blacktop. The road ran between the houses and the river, but the riverfront was privately owned anyway. Little divots of lawn cowered between pavement and riprapped riverbank, each with a small dock and enough "No Trespassing" signs to supply every one of the Thousand Islands. At the end of the road a small parking lot snuggled below the low bluff holding the 144-year-old stone lighthouse. We got out into what had become a light, intermittent rain and stood on the shingled shore, watching Lake Ontario slide imperceptibly into the St. Lawrence River.

This is where the Great Lakes end. Drops of water that spent five hundred years just finding their way out of Lake Superior mingle with drops falling as rain right now off Cape Vincent, and together they slip gracefully down the river, threading the Thousand Islands, sniffing their way toward the distant, mythological Atlantic. Hydrologically speaking, what I have just written is a lie: there is no detectable current in the St. Lawrence until it falls over the Moses-Saunders Dam, near Cornwall, Ontario, a hundred miles downstream. The Lake Ontario regulatory works are there, run jointly by the U.S. Army Corps of Engineers and the Canadian Coast Guard through a bilateral commission known as the International St. Lawrence River Board of Control. Operating under Plan 1958-D—the regulatory regime Bob Beltran had complained about so vociferously back in Chicago—the Moses-Saunders Dam keeps Lake Ontario's level inside a four-foot range, from 243.3 to 247.3 feet above sea level. This means that the upper river is really just a long, skinny extension of the Lake. But hydrology and psychology part company at Tibbets Point, where the lighthouse stands by the head of the island-spangled channel and looks toward the western horizon over the long waves of the infinite sea. Despite all the development on the way out there, the place still feels lonely and primeval, little different than it must have felt to Samuel de Champlain, paddling past on his way home to Montreal nearly four hundred years ago.

Back in the town of Cape Vincent we turned south on 12E. The rain had stopped entirely, and the early-evening sun was peeping fitfully through tattered and dissolving clouds. At Chaumont, named for a French baron who settled here after befriending Ben Franklin and helping broker the Treaty of Paris, we turned left, onto Morris Track Road. We were seeking the Chaumont Barrens, an odd and ancient landscape with few counterparts anywhere else on the planet.

Ten thousand years or so ago—according to the best guesses of the scientists who have studied the place—an ice dam broke somewhere along the edge of the retreating continental glacier to the north of Lake Ontario, suddenly releasing several hundred cubic miles of meltwater. The resulting flood swept into the Ontario basin and out again down the St. Lawrence, right over the top of Cape Vincent. Much of the soil was swept away. On most of the cape it built right back up again. On the land that became the Chaumont Barrens, it did not. The Barrens are a flat limestone plain which had given under the weight of the glacier, creating a pattern of right-angled fractures, and now that the ice had gone the plain was slowly rising, expanding the fractures into narrow cracks. When the flood came through, the cracks were exposed. Solutional processes widened them further. With holes gaping every few feet, the plain would not hold water. A few plants gained precarious toeholds, so a little soil built up, but the process was extremely slow. It remains slow today. Ecologically speaking, the Chaumont Barrens are laggards. The lands around them have evolved to mixed hardwood forests. The vegetation of the Barrens remains much as it was right after the ice left. There simply has not been enough time, at the pace of things here, for plant succession to operate.

Asters and spruce at Chaumont Barrens near the Thousand Islands,
New York

Lands with this odd combination of geologic and ecologic history
are called *alvars*. All that are known, worldwide, lie in two clusters.
One of these clusters is in Scandinavia, where their peculiar characteristics were first noted and named. The other is here in the Great Lakes.
Alvars lie at scattered locations along the northeastern quadrant of the
Niagaran Cuesta, from Ontario's Bruce Peninsula northward across
Manitoulin Island to northern Michigan's Drummond Island. There
are also a few tiny spots with alvar-like characteristics at the north
end of Green Bay, and a few more on the Bass Islands. And there are
the Chaumont Barrens.

I was unaware of the Chaumont Barrens in 1983, but that is not
surprising: no one else was aware of them, either. It was not until
the late 1980s, when an ecologist from The Nature Conservancy
became intrigued by some regular linear patterns of vegetation she
had spotted in aerial photos of the Cape Vincent peninsula, that
anyone with the training to see what was here paid them a visit. Today,

the Conservancy owns the place, or at least 1,633 acres of it. There is a two-mile-long nature trail—opened in 1995—an interpretive kiosk, and several benches. There are still not very many people. The various "Guides to the 1000 Islands" we picked up at stores and Chamber of Commerce tourist information centers around the cape do not mention the Barrens at all. If this is a shame, though, it is a delightful one. It pretty well guarantees that those who *do* find their way here will have it mostly to themselves.

In dusky light, through air cooled by the recent rain, we walked the Barrens Trail. The flora was almost Arctic—low junipers, mosses, and lots of *Compositae* as well as the sparse grasses. There was goldenrod; there were scores of daisies. The few trees were largely spruce and cedar. Underfoot the limestone plain was crisscrossed by the solutional cracks that make the place possible. A chickadee called. It was though we had stumbled across a bit of the taiga, the forest of the far North, all alone and unpeopled in the midst of the Thousand Islands. A soul-reconstructing place. I hope New Yorkers always protect it and never find it.

Afterward we drove to Clayton, had supper in a tiny hamburger stand across from the local high school, and came back to the motel. We tried leaving the windows open as we had in Webster, but a phantasm of tiny, almost invisible gnats found their way into the room *through* the window screens. I closed the windows and turned on the A/C. The next day we would turn the corner and head west.

VI
THE BRUCE, AND BEYOND

THE INTERNATIONAL BRIDGE WAS ACTU-
ally several bridges, leapfrogging from island to island across the
broad upper end of the St. Lawrence. The polished Canadian Shield
granite gleamed nearly white; the blue river sparkled in the morning
sun. On the Canadian side, Canada 401 had been moved since 1983
and was now a freeway several miles back from the water: the river-
edge route Larry and I drove had been rechristened the Thousand
Islands Parkway. I went by memory—which I should not have done—
and guided us onto the freeway, which I should not have done, either.
SUVs humped past; BMWs flew by like bombs, their one-handed
drivers cradling cell phones. The speed-limit signs said 90 km/h—
about 56 mph—but people seemed either to be ignoring them or
reading them without the "k." The primary culture shock for a U.S.
citizen going to Canada is that there *is* no culture shock. We are two
nations, but we are one people. If you want to visit a foreign land,
you will have to go much further afield than this.

At the tiny town of Marysville we escaped down Highway 49
toward Quinte's Isle. The rush of the freeway gave way to rural two-
lane pavement. With Melody driving, we crossed the high-arched
bridge over the Bay of Quinte—built to clear the masts of sailboats
on the popular Rideau-Trent-Severn Waterway—and descended onto
the green island.

Quinte's Isle was actually a peninsula until a little over a hundred
years ago. An irregularly shaped mass of land forty miles long by
twenty-five wide, it extends into the east end of Lake Ontario like
an inkblot spreading from the otherwise cleanly drafted north coast.
Between it and the mainland zigzags the Bay of Quinte, sixty miles
long but never more than six miles wide: at the bay's head there is a
tiny, low isthmus scarcely a mile across, and then the Lake starts up
again. The isthmus, for easily understood reasons, is known as the
Carrying Place. When the Murray Canal was cut through in 1889—

replacing the wooden railway over which oxen had been pulling bateaux and barges since the beginning of commerce on the Lakes— the carrying ceased, and the Isle became an island in truth as well as in name.

There was a Tyendinaga tribal village, called Kente, at the Carrying Place, and that is where the bay's name—and eventually, the island's—came from. The island was originally called Prince Edward, but that caused too much confusion with Canada's other Prince Edward Island—the one in the Atlantic, off Nova Scotia—so the name was rapidly dropped, although it is still attached to the map here in two places: Prince Edward Point, the island's southeast tip, and Prince Edward County, which is coterminous with Quinte's Isle. This is Loyalist country. After the American Revolution there were still colonists in the thirteen breakaway states who remained loyal to the Crown, and quite a large number of them emigrated to Canada and were given land grants here. The 390 square miles of the island were thus settled in a pattern strikingly reminiscent of late-eighteenth-century New York and Pennsylvania. Much of that pattern remains intact today. Small villages and farmsteads scatter over the land; fields alternate with woodlots, punctuated by tidy brick or white frame buildings. An air of simpler times rests easily here. Picton, Prince Edward County's metropolis, weighs in today at less than 4,400 people.

But change is coming. Most of it is taking place along the island's deeply indented, highly scenic coast, where the Lakes' current scourge, suburban sprawl, has made major inroads. It is not readily apparent from the main highway corridor, which stays inland for most of its length, but it is there. Quinte's Isle's population is currently growing at a rate exceeding 11 percent per annum. Ranch homes are swallowing lakeside pastures. West of Wellington we stopped at the island's only roadside rest area—a small, shaded green lawn with picnic tables eyeing Lake Ontario across two hundred yards of fenced pasture. The view of the Lake is splendid. It is also doomed. The eastern part of the pasture, active farmland when I saw it in 1983, was full of houses by 1998. The western part would soon follow. The most prominent sight here this year was neither the Lake nor the pasture, but the large "For Sale" sign right in the middle of the rest area's already constricted view. The landowner has a legal right to put houses there, of course, but it is a sociopathic step nonetheless. In the name of property rights we are reducing views to vulgarity and parks to porta-potty sites. I have had quite enough of the "takings" movement. We need a "givings" movement—one which will give our North American birthrights of open space and access to nature back to us.

While we sat gloomily contemplating the real estate sign and the porta-potty, a cheery codger on an old one-speed bicycle came rattling into the parking lot and disappeared into one of the objects of our contemplation. When he came out again he stopped by our table. He looked like a fishing bum, and because he was a retired civil servant I guess you could say that was a pretty good description. "What brings a couple of Oregonians to Quinte's Isle?" he asked.

"Just passing through," I said.

"Is this your first visit?"

I explained about the 1983 trip and reasons for the current one.

"Well, change is change," he said. "It's not good or bad, it's just change." He pointed west along the road. "I've got a little place up

The view of Lake Ontario from the rest area west of Wellington on the Isle of Quinte, Ontario

there. I go fishing a lot. Nobody bothers me, and I don't bother anybody. I've got a lot less view than when I came, but a lot more nice neighbors. It weighs out."

"You seem pretty happy."

"I am. I go fishing. I get into town on this thing every couple of days." He slapped the bicycle. "I've got a car, but you don't really need one here. I don't use it but once in a while." His stubbly face turned from one of us to the other, beaming beatifically. "Well, I'd best be getting on. You folks enjoy your stay."

He straddled the bicycle and rattled away toward Wellington. The whole time he was there the smile had never left his face.

Back on the mainland we were swept into the maelstrom of Canada 401 and whirled toward the Toronto Vortex. We tried to escape to the Lake. The first try found a wetland, but beyond it only blind alleys toward the shore. The second time got as far as a tiny information booth at the freeway entrance, where the fresh-faced blond behind the desk shook her head when I asked if her town had beach access. "We do," she said, "but it's only about as wide as this room. There's a nice park in Cobourg. You should go there."

The third try was in Cobourg. We didn't find the nice park; we found more traffic and more dead ends. We had a campsite waiting on Georgian Bay, several hours away to the northwest. I had wanted a farewell tryst with Lake Ontario, but it was not to be. We slipped back onto the freeway and abandoned ourselves to the westward flow.

I don't know what the upper limit of population for a functioning human society is, but I do know that Greater Toronto exceeded it long ago. The city is a whirl of people constantly crossing each other's paths, like neutrinos in a cyclotron. The pace and volume of traffic, the raw, wounded-looking land beneath the new subdivisions and shopping centers, and the general sense of noise and hurry made me feel thoroughly jangled, like a mouse caught in a bagpipe. We turned north on 404 well before the city center, trying to escape the worst of it, and maybe we did: but if so, I would not want to see the worst. The freeway ended and we found ourselves facing traffic lights. Cars shuffled and crushed between them; trucks roared and fumed. Everywhere there was new construction—new homes, new strip malls, new office buildings. This, John Hartig had told us, was the finest agricultural land in Ontario. That has not stopped people from paving it. Hundreds of thousands of acres of it. The property-rights people have clearly won the war here. They can go to bed and sleep well. I wonder what they are planning to eat.

We got back on the freeway, booming north, away from the hell they are building out of what once were Toronto suburbs. Traffic thinned; the countryside began to open up. Far to the west a thin blue line on the land came closer, resolving into a long, even range of hills. Once again we were approaching the Niagaran Cuesta, here called The Mountain. The Mountain begins at Niagara Falls and sweeps north across what is known to travel agents as Huronia. It strikes the south shore of Lake Huron and keeps going, thrusting the long thin finger of the Bruce Peninsula northward almost to the Lake's center. From there it continues as a chain of islands, arcing north and west all the way to distant Michigan. Back at the beginning of the trip, coming in at the Door, we had encountered Niagaran dolomite. We had seen it in southern Wisconsin, spent the night on a spur of it in the Bass Islands, and dropped over its edge east of Buffalo. Now here it was again, binding the trip, and much of the Lakes, within an immense circle of blue-white stone. Four hundred million years ago, billions of tiny diatoms laid down their lives in the shallows of the Silurian sea and created, all unknowing, one of the great scenic landscapes of North America. Humans are not the only creatures whose actions significantly alter the planetary surface.

As we approached Barrie, signs pointing to Collingwood shunted us off the freeway onto a set of small secondary roads. I'm sure the Ontario highway department's motives were noble—saving tourists the trouble of going through Barrie, saving Barrie the trouble of dealing with tourists—but the results were not exactly satisfactory. The route was all two-lane, and it was complex. Intersections were well signed—in marked contrast to Buffalo—but there were far too many of them. We were shifted west, north, west again. Heavy equipment shuttled slowly along the roadway, backing up long skeins of traffic. At one point a large stone wall loomed on the right, and behind it we could see the roofs of yet another brand-new suburb: the backside of Barrie, spreading west.

Finally we reached Collingwood. Trim houses nestled along tree-shaded streets: a compact business core of clean brick buildings spread back from the harbor. We made a brief search for a harbor view and found a wetland at the end of a gravel parking lot behind a gas station. The wetland was for sale. I was suddenly overwhelmed by a desire to see a shoreline without a "For Sale" sign on it. We postponed the harbor search to the next day and hurried west along Georgian Bay toward Craigleith Provincial Park.

Geographers sometimes refer to Georgian Bay as the "sixth Great

Lake." There are good reasons for this. One of them is size: at 120 miles long and more than 50 wide, this "bay" is so big that if it were a separate Lake it would rank among the twenty largest on the planet. Another is degree of separation: barricaded behind the Niagaran Cuesta, the great bay functions as an almost completely separate system, with only modest amounts of interchange with the rest of Lake Huron. The biggest reason, though—at least to a casual observer—is character. Georgian Bay has far more in common with Lake Superior than it does with Lake Huron. Like Superior, it is cupped in rock. The Canadian Shield dips into the bay on the north and east, forming the Thirty Thousand Islands; the Niagaran Cuesta rises like a wall to the west. The bay's cold water has the crystalline clarity that comes only with a rocky bottom and a stony shore. With zebra mussels or without, Lake Erie will never come close.

At Craigleith we pulled off the bayside road. Soon we had the tent set up, looking through woods toward water from the edge of a small knoll. Behind us were maples and grass, golden in the sun; then the highway and, beyond it, the soaring face of The Mountain. Here at Collingwood, at the highest point on its circuit, the Niagaran Cuesta deserves its nickname. From bay to crown the rise is twelve hundred precipitous feet. Bathed in blue evening shadow behind the sunlit maple savannah of the campground, it had a brooding look, like an advancing earthen storm.

We wandered down to the water. Level ledges of gray mudstone stairstepped far out into Nottawasaga Bay, the twenty-mile-wide bight into the south shore of Georgian Bay on which Craigleith and Collingwood are both located; the surf swept over the shoremost shelf, leaving pools of clear water cupped in hollows. The black hills of Cape Rich, to the west, shouldered the sinking sun. The pools turned golden. Campers gathered, their conversation hushed. Gulls pedestaled on small boulders, looking west, as we did, toward the dark water and the flaming sky. They, too, were still. I wonder why we think animals lack an aesthetic sense.

Back in downtown Collingwood the next morning we went looking for a perfect-ten breakfast and found a good 9.5. Then we drove straight north and came out on the edge of Collingwood Harbour, at Harbourview Park. The glassy water held geese and reflections. A young towhee cultivated a trailside thicket. Overhead, thin wisps of clouds wavered and dissolved in the brightening air.

At one corner of the park, cattails and lily pads graced a small wetland. This marsh at the mouth of the old Oak Street Canal—and a

Shoreline development and Canada geese at Collingwood Harbour, Collingwood, Ontario

similar but much larger one at the mouth of Black Ash Creek, west of the park—are major parts of the Collingwood story. Though they do not look it, both are artificial, created by community effort in the early 1990s. Their contribution to wildlife habitat improvement is obvious. Their role in solving the harbor's nutrient enrichment problem may be less apparent, but it is no less real. Like all wetlands, they polish the water that passes through them, removing both particulates and nutrients. For Black Ash Creek, the primary sources of these are nonpoint—agricultural runoff, drainage from lawns and streets. For the Oak Street Canal, the primary source is the Collingwood Sewage Treatment Plant.

There has been a great deal more than just wetland construction involved in the delisting of the Collingwood AOC, of course. Water conservation measures—low-flow showerheads, toilet dams, and the installation of water meters—have been implemented, reducing the load on the treatment plant by 35 percent: this has cut phosphorus loadings to the harbor by a similar amount. A new high-tech treatment system went on line in 1993. The city's land-use plan calls for all new development on the waterfront to not just preserve remaining wetland habitat, but increase it. Eight thousand cubic meters of contaminated sediments have been removed from the harbor. After Christmas one year, the community collected castoff Christmas trees and encased their bases in concrete. They were placed on the harbor ice. When the ice melted in the spring the trees went to the bottom, where their branches formed excellent shelter for breeding fish. That year the harbor's fish production went up a remarkable 1,000 percent.

Holding it all together has been an extremely high level of citizen involvement. The water-conservation program mentioned above reached 70 percent of Collingwood's households; the Christmas-tree reef collected a thousand trees from a city of barely fifteen thousand people. The twenty-four-member Collingwood Stakeholder Group— the RAP's Public Involvement Committee—includes an unusually broad cross section of the community: developers, environmentalists, city and county government representatives, the Chamber of Commerce, service clubs, motels, the local library. Work parties are held; grants are sought, won, and managed by the group, which shares its work with the community through a newsletter called *The Daily RAP Sheet*. No other RAP, to the best of my knowledge, has achieved Collingwood's level of commitment and coordination. That is probably a principal reason why no other RAP has been delisted. Collingwood achieved delisting in November 1994.

It would be an error to state that delisting means all will be well in Collingwood forever after. Vigilance remains a necessity. The new wetlands have been invaded by purple loosestrife: weeding parties organized by the Stakeholder Group have made inroads into the enemy's ranks but have failed to eradicate it. We saw loosestrife in the Oak Street Canal wetland the day of our visit. The town itself has also suffered an invasion. One of the loveliest small cities in Canada, it has begun to draw not only tourists but new residents. On the far side of the harbor from the path we walked in Harbourview Park, beyond the geese Rorschached by the mirror-smooth water, a large new development hulked up from the shore. Because of the strictures that have been placed in the city's planning ordinances, it is reasonably certain that this development did no damage to wetlands or waterside habitat. This does not mean it was without impact. Increased numbers of houses and paved streets mean, at the very least, changes in the pattern in which rainfall infiltrates the soil. Increased sewer connections mean increased sewage, biting into the gains made by the water-conservation program. And even if the infiltration problem can be remediated and the sewage increase can be absorbed by the plant, we are dealing with new residents here. They have less investment in the life of the community, and therefore less reason to support the efforts put forth by the Stakeholder Group. In fact, since it is part of the local establishment, they are likely to see it as a "good old boy" network and distrust it. This is the flip side of the positive new-resident effect I pointed out back on the south shore of Lake Erie, but it comes from the same place: lack of familiarity with the local environment. I have no desire to impugn the motives of Collingwood's newcomers. I only wish to point out what being newcomers means.

Collingwood's spanking-new sewage treatment plant borders Harbourview Park to the east. It is odorless, noiseless, and visually unobtrusive—a tribute to the art of sanitary engineering. If you wish a reminder that this in itself will not permanently solve the problem, though, you need look no further than the treatment plant's own parking lot. There, beneath a large, spreading hardwood tree, squats a bright blue porta-potty.

No solution can deal with more of a problem than it can reach.

We climbed The Mountain by car, a thousand feet up the northeast face via the Scenic Caves road to a viewpoint a mile or so beyond the caves. Collingwood was a toy town from that height, spread out around a thimbleful of harbor. Beyond it Georgian Bay spread, flat and blue, to a horizon so distant it appeared curved, and there was nothing toylike about that part of the view at all.

Back at water level we turned northwest. The Mountain lifted to our left; the bay lay to the right. Tiny towns—Craigleith, Thornbury, Meaford—came and went. The road lifted itself west over Cape Rich and came down into Owen Sound, poured picturesquely over the steep hillsides surrounding the head of its narrow bay. The city seemed squeezed by its setting. Buildings jostled against narrow streets which joined each other at odd angles. Traffic bunched and growled. We lost the way and had to double back. Finally on the far side, we found Highway 6 and turned north, up the spine of the Bruce Peninsula. The hustle of Owen Sound thinned and evaporated: farms and orchards spread to either side. The road bored through woodlands. Signs pointing both east and west bore nautical names, but there were no views of the water.

The Bruce is where The Mountain goes to sea. The Niagaran Cuesta, reduced somewhat from its height at Collingwood but still impressively tall, strikes the south shore of the world's fifth-largest body of fresh water and simply keeps going, thrusting a long spine of dolomite toward the cold heart of Lake Huron. On a map it is a mirror of the Door, and I had expected it to share the character of its Wisconsin twin. This assumption is correct in general, but I was not prepared for how far it would vary in particulars. The Bruce is higher than the Door, and wilder, and more rugged. The water surrounding it is broader and colder and deeper. Though it is actually slightly further south, it feels much further north. Northern forests—birch, maple, cedar, spruce—stand against light gray stone. Farmers scratch, rather than till, the thin soils. The eastern shoreline is deeply indented, forming a series of long, narrow bays. Small villages, showing their fishing roots, cling to the heads of the bays beneath near-vertical dolomite cliffs.

Near the tip of the peninsula, Bruce Peninsula National Park sprawls over fifty square miles of Cabot Head. We drove the Emmett Lake Road to its end and hiked the half-mile trail to the Halfway Log Dump. In the peninsula's not-so-dim logging past, trees would be felled in the winter and skidded over the snow to the Georgian Bay coastline, where they would be stored—"dumped"—until Lake transport opened up in the spring. Hence this unglamorous name for one of the most striking places on the Great Lakes. The flower-lined trail followed the old skid track through small sunny openings and dark second-growth forest. At its lakeward end it dipped abruptly and broke out onto a steep beach of angular, blindingly white pebbles. To the left the beach ended against a tall headland: to the right it kept going, a long slim crescent of white bending south beside blue water that stretched beyond the horizon. Behind the beach, green forest swept steeply up to near-vertical stone. These cliffside forests were passed over by the early loggers as too difficult to cut, and some of the gnarled white cedars that grace them are more than a thousand years old—the oldest trees in all of eastern North America.

We wandered down to the water. Little clear waves washed in like ghosts of surf; stones glistened on the sunlit bottom. The white pebbles turned gray when moist. Had it not been for that, and for the abrupt change in refractive index, it would have been impossible to tell where air stopped and water started.

In Tobermory that evening, at the very tip of the Bruce, we strolled to supper along Little Tub Harbour. The water in the harbor was so clear we could see the undersides of the hulls of boats pulled up to the concrete dock. In a tiny restaurant perhaps a hundred feet from the water's edge we ordered fish and chips. The fish was Lake Huron whitefish, and only hours earlier it had been swimming around in Lake Huron. We meandered back to our motel. The balcony in front of the room overlooked the harbor, which was full of fishing boats. Beyond the low, rocky arm of land separating Little Tub from Big Tub we watched the great white bulk of the M.S. *Chi-Cheemaun*—the ferry that connects the Bruce Peninsula to Manitoulin Island—glide slowly into its slip, in sun though most of the village was now in shadow.

The beach at Halfway Log Dump, Bruce Peninsula National Park, Ontario

Its sleek bow opened like a clamshell. Cars poured forth, their lights turned on against the coming dusk.

While this was going on, a fiddle started playing in the distance. We followed the sound to the docks. The fiddle player sat on a small balcony on the back of a large sailing yacht, playing an electric fiddle which was totally bodiless. Melody, a fiddle player herself, was intrigued. A blond woman pounded out chords on a keyboard; a second man played spoons. The tunes were Gaelic—jigs, reels, and hornpipes—but with a Western twist which indicated they had probably been filtered through one of the North American fiddle traditions. Eventually the fiddler broke off, bowed to the audience that had gathered, and disappeared into his boat. We walked back to the motel. The ferry was just disappearing into the distance, lit by the last rays of the sun. The town turned rosy; the sliver of Georgian Bay visible from the motel polished itself silver. The lights of the village clung to the limestone hills and reflected in the harbor. Who needs Greece?

Stones beneath two feet of water at Halfway Log Dump

The Great Lakes are often referred to as the American Mediterranean: the Bruce Peninsula is a place where this description rings true. White limestone rises from blue water; masts dip and sway in snug harbors. There is heavy tourist development in the small bay towns, but there are few of the subdivisions that plague the rest of the Lakes, and no strip malls, fast-food restaurants, or big-box stores. Out at the tip, Tobermory feels isolated, almost insular, as if the connection to land at the base of the peninsula didn't exist. I hope it can stay that way, but, given its amenities, I fear very much for its future.

THOSE SEEKING TO CATCH THE *CHI-Cheemaun's* 7 A.M. run to Manitoulin Island are asked to be in line at the ferry terminal by 6 A.M. This is easier for some than for others. Melody, who is the day person in our family, was up almost before the alarm stopped ringing at 5:15; she grabbed a shower and had most of our luggage back in the car before I staggered out of bed at 5:45, "up" only in the sense that my posture was semi-vertical. At the dock the boat was waiting, its clamshell bow raised, its cavernous interior brightly lit in the blue predawn light. There was an hour to go before sailing. We left the car in line and slipped into a dockside coffee shop.

The ship got under way right on schedule, backing away from its slip into Big Tub Harbour, hydraulics groaning as they lowered the bow smoothly into place. Thus set, the *Chi-Cheemaun* looks like a cruise ship: long, white, and sleek, with a rakish stack above two stories of cabins. "Chi-cheemaun" means "big canoe" in Ojibwa, and at 340 feet in length, with a crew of 36 and a capacity of 140 cars and more than 600 passengers, the boat certainly qualifies as big. The run to Manitoulin across the island-spangled sea of the Main Channel took an hour and a half. Golden mist lay on blue water, the rocky islands black against it like shadows of ancient myths.

In the forward lounge stood a small computer, its screen displaying a detailed map of a portion of the Main Channel. A red line and a blue line, almost parallel, slashed diagonally across from upper left to lower right; where they met in the middle there was a small circle which jerked forward every few seconds. A set of numbers at the bottom of the screen shifted constantly.

"It's gotta be a GPS," said a voice slightly above and behind my left ear.

I turned. The man was tall and slender and sixtyish, and the mild eyes behind his spectacles gleamed with boyish enthusiasm as he studied the screen. "I think the red line's our course," he said.

"Probably," I agreed.

"Have you figured out what the blue line is?"

"I think it's the ship's heading."

He nodded. "And the numbers must be the soundings." He grinned and stuck out his hand. "I'm Ted Sobel," he said. "Sorry for the intrusion, but I figured anyone studying the global positioning system must be worth talking to. Isn't it fascinating? Where is everybody?" He looked around with a mystified expression. Indeed, Ted and I were the only ones I observed to do more than glance at the GPS screen during the whole crossing.

Ted was a retired professor of agricultural engineering from Cornell University; he and his wife were on their way to visit a daughter in Michigan. I asked if she lived in the Upper Peninsula. He shook his head. "No," he said, smiling. "She's at an environmental studies center in the Lower Peninsula. We're taking the circle tour."

"Lansing?"

"The Jordan Valley. The Au Sable Institute of Environmental Studies, in Mancelona."

"I'm not familiar with that one."

"Not very many people are. It's run by a consortium of Christian colleges. The idea is to encourage the stewardship of God's creation."

"Sounds like a good idea to me."

"I thought so, too. I don't know why we always leave taking care of God's handiwork to the atheists."

He handed me his card. It read:

THE WHITE CHURCH CABIN COUNTRY STORE
Items of Old-Fashioned Practical
Value. Week-ends by chance or call
xxx-xxx-xxxx.
Ted Sobel, Proprietor.
'Seek ye first the kingdom of God,' Matt. 6:33

"Just a little business to keep me out of trouble in retirement," he smiled. "Drop by if you're ever in Ithaca."

Landfall at South Baymouth, on Manitoulin, was coming up shortly. I excused myself to find Melody, and I didn't see Ted again.

THE BRUCE, AND BEYOND

I put his card away carefully, though, and during the rest of the trip I often pulled it out and looked at it. "Week-ends by chance." In this frenetic world that is a truly radical outlook. Is it necessary to point out that its author was the most blissfully peaceful person I encountered on the entire trip?

Perhaps it was the comparison to the day before, or perhaps it was the dull reality of land after the splendid fantasy of dawnlit sea over which the ferry glided. Or perhaps it was the fact that we had been forced to get up so bloody early. Whatever the cause, Manitoulin Island was a bit of a letdown. Our time on the island went through all the right motions, but it never reached any real emotional highs.

The island was certainly attractive enough. Long and irregular, accented by water and forest, this sea-isolated section of the Niagaran Cuesta is the largest freshwater island on the planet. Deep bays scallop far back into its shores. Most communities are tiny, a scatter of buildings along a couple of crossed streets. There is good public access to the water—an unfortunate rarity in the Great Lakes Basin, especially on the Canadian side. There are waterfalls, and inland lakes, and even a couple of good-sized rivers. There are high viewpoints that look out over the sparkling North Channel to the rocky La Cloche Mountains on the distant mainland. It is beautiful, but it doesn't quite add up. Whatever it has, the Bruce seems to have more of. It is Manitoulin that is really an island, but it is the Bruce that feels like one.

At the town of Little Current we walked a sunny waterfront boardwalk lined with sailboats, many of them flying United States colors. At Ten Mile Point we perched on the edge of the Cuesta and ate peaches and peanut butter sandwiches, nourished as much by the big North Channel views as by the food. At Kagawong we bypassed the jam-packed parking area at the brink of Bridal Veil Falls and drove another half mile downhill into town, where an empty parking lot and an equally empty trail along the Kagawong River beckoned; the trail led past an old power plant and through shady woods full of cardinal flowers back to the base of the falls, a sixty-foot curtain of falling water leaping from an overhanging limestone ledge, its splash pool laughing with swimsuit-clad families who had come in the cars parked at the rim.

Around 2 P.M. we came out on the south shore of the island at Providence Bay. Low barrier dunes stretched behind a mile-long crescent of sand; a new boardwalk ran lengthwise along the crest of the dunes—we are protecting sand well these days—and there was a modern, strikingly designed visitor center with museum exhibits, craft stores, and a small café serving what I will rashly state is the best ice cream in Ontario. I waded off the beach into the clear water through floating bits of broken macrophytes, small surf slapping at my knees, looking out the bay's mouth into the open Lake. When Manitoulin Island was settled in the mid-nineteenth century, three skeletons were found buried beneath boulders near this bay, their bones covered by remnants of decaying sailcloth. Nothing has ever been learned of their origin. Were they part of the crew of La Salle's ill-fated *Griffon,* the first sailing ship on the Great Lakes, sent off from Green Bay with a cargo of furs and never seen again? The grave of the *Griffon* has never been located, despite several centuries of attempts. Its course would have taken it very near here. Do its wooden bones lie buried in the sandy bottom of Providence Bay? As far as I know, that tantalizing possibility has never been properly investigated.

We were off Manitoulin by 4 P.M., hopping from island to mainland across winding blue channels once plied by voyageurs' canoes. Manitoulin is made of Niagaran dolomite; the mainland is Canadian Shield, but with a difference. Where most of the Shield is granite, the section just north of the North Channel is quartzite. Under the erosive power of Precambrian weather, before there were plants to protect the planetary surface, large quantities of granite sands were created and carried into a shallow sea south of the main body of the Shield, where they became sandstone. Later—but still in Precambrian times, approximately two billion years ago—the buried sandstones were suddenly squeezed, twisted, and cooked. The process seems to have been related to the creation of the Sudbury Basin, thirty miles to the northeast. Geologists are cautiously vague about any event that ancient, but the evidence strongly suggests extraterrestrial influence. The earth appears to have collided with an asteroid. It was six miles across, and it buried itself in the stone of the Shield, leaving a shallow basin sixty miles wide. Evidence of the impact is detectable as much as nine miles below the surface. The nickel-rich asteroid is still there: the miners of Sudbury have been making a living from it since 1884.

Whatever their source, these North Channel quartzites are incredibly beautiful. The chemical changes accompanying their formation left them hardened and shot through with a broad range of pale hues—reds, whites, and grays, mostly, but with some blues and yellows and even a few greens. Black flecks of biotite mica accent them. Shaped and polished by the glaciers, they form little steep mountains of mostly bare stone. Small clear lakes pool on the stone, dotted with smooth islands; rivers cascade down them. Geographers call this region the La Cloche Mountains; travel brochures call it Rainbow Country. For me, Paradise will do. Sitting beside one of the little roadside lakes, on a finger of smooth red stone ending in lapping blue water leading off among cliffs and islands, I found myself quoting the line about Washington State's Ramparts Ridge made famous by Louise Marshall: *If this isn't heaven, when my time comes I'll refuse to go.*

On the far side of the little range of mountains, in the early evening, we came down into Espanola. There was another clear lake, called, imaginatively, Clear Lake; our motel faced it across a small park. In the gathering dusk, with all things either gray or black, we walked outcrops of star-contorted quartzite molded and marked by two miles of ancient ice. The quiet water stretched west, a flat brightness between dark shoulders of trees and rock. Lights twinkled among the trees: houses, mostly new, most of the way around the lake. They were unobtrusive but disturbing, more for what they meant than what they were. After giving us two days of relative freedom, sprawl had shown its hand again.

There is not nearly so much development up here as in the lower Lakes: The Line appears to be holding. But there are signs of change. Large blocks of land, zoned for development, were for sale near Little Current. Espanola has developed the fast-food outlets and big-box stores that the Bruce and Manitoulin have so far staved off. There are no freeways, but this is freeway culture nonetheless, and its ugly future is not hard to predict.

ESPANOLA BORDERS THE SPANISH RIVER, named for the Spanish-speaking wife of an early voyageur. On our way out of town in the gray morning we stopped to look at the river, running smooth and shining over stone ledges. Thickets of willow bordered it beneath a tall bluff of dark quartzite. Just upstream from the Route 6 bridge, on the south bank, squatted an E. B. Eddy, Ltd., paper plant. The plant was spread out across the landscape in several large mechanical-looking chunks, as if someone had dropped the *Chi-Cheemaun* onto the riverbank from a great height. White steam rose from it into the already steamy air.

In 1983, this plant was a pariah. On July 9 of that year, as Larry Chitwood and I stood on the Door Peninsula gawking at our first big view of the freshwater ocean, 47,000 gallons of "soap"—a halide-rich by-product of paper manufacturing—were spreading out of the mouth of the Spanish River into the North Channel, and an estimated 100,000 fish were going belly-up. Angry anglers filled pickup beds with dead fish and drove to Espanola, where they dumped them in front of the plant office. "We wanted the staff to have a look and have a whiff," one of them told *MacLeans* magazine.

The halide spill was not an isolated incident: the mill had been dumping junk in the Spanish River since its first primitive predecessor started grinding out wood pulp in 1905. Paper production began at Espanola in 1912; a kraft mill, papermaking's messiest operation, was built on the site in 1943. By the late 1970s the plant's assault had turned the bottom thirty-two miles of the river into a virtual desert. Benthic fauna—bottom-dwelling invertebrates—had declined to below detectable levels in the riverbed: walleye populations were severely reduced in the river's mouth and in adjacent parts of the North Channel, and channel catfish and redhorse suckers were gone. The few fish that remained tasted funny. These were among the reasons given when the International Joint Commission declared the river and harbor an Area of Concern in 1980—one of only four to be established on Lake Huron.

The timing was ironic, because the solution had already arrived. Although the 1983 spill was not the first problem the Eddy plant had caused for the Spanish River, it would be among the last. The key event was Eddy's 1969 purchase of the plant from a Michigan firm called Kalamazoo Vegetable Parchment. Just about the first thing Eddy did was to install primary effluent treatment for the antiquated facility, which had simply been letting the paper-process wastes run into the river. In 1977 the plant's new owners converted the softwood delignifier—the processor that removes the hard parts of the wood prior to papermaking—from a chlorine-based to an oxygen-based system, the first North American plant to do so. By the time the AOC was declared the hardwood delignifier had also been converted, reducing the plant's chlorine usage by 50 percent.

In 1981 Eddy began construction of secondary sewage treatment lagoons at the Espanola plant; and in August 1983, with little of the fanfare that had accompanied the spill of "soap" a month earlier, the new facility came on line. Improvement was immediate. By 1987, when the Annex to the Great Lakes Water Quality Agreement was signed and the Spanish River AOC was carried over into the new RAP process, the river's recovery was already mostly complete. Walleye, channel cats, and suckers had returned to the harbor, and benthic fauna counts had climbed from zero back up to more than eight thousand per cubic meter. Angler "taste tests" conducted in early 1987 found no remaining off-flavors in the fish. With the fouling stopped, the fouled area had bounded back. The recovery powers of nature are amazing. All that is necessary is to give them a chance to work.

The paper mill's commitment to improvement hasn't stopped with secondary treatment of their effluent. As Melody and I stood beside the Spanish River in 1998 the plant was in the middle of a three-year, $93 million upgrade to its papermaking machinery, with a goal of totally chlorine-free paper production by 1999. The company expected to be the first in Canada to reach that objective. The Spanish River AOC has not yet been delisted as I write this in early 1999, but delisting appears imminent. If it occurs by the end of this year, half of Lake Huron's AOCs will have been declared clean before the millennium, compared to precisely none elsewhere in the Great Lakes Basin. Huron may be, as many have termed it, the "forgotten Great Lake," but that is only by outsiders; Lake Huron people care for their own.

From Espanola we drove west along the Trans-Canada Highway under a sky wiping itself clean of clouds. The highway followed the north side of the North Channel, staying well back from the water. Little towns came and went: Massey, Spanish, Algoma Mills, Blind River. Hardscrabble farms scratched among conifers. A sign said *Cozy food. Warm beer. Cold rooms.*

In 1878 or thereabout, a young man named James Whalen—or perhaps James Phelan—was killed while trying to break up a logjam at a waterfall called The Chutes on the Riviere Aux Sables, and a song was made about him. The Chutes is now an Ontario provincial park, and we stopped there in midmorning. The tannin-rich water thundered over a broken ledge of Canadian Shield granite, vibrating the wooden viewing platform at the brink. The riverside trails were empty. A short distance west a small rural road called Whalen Street intercepted the highway. Primed by the park, we immediately burst into song:

> Oh Jimmy, oh why can't you tarry here by me?
> Don't leave me alone, so distracted in pain.
> Death is the dagger that has drove us asunder;
> Wide is the gulf, love, between you and I.

For three days—ever since leaving Lake Ontario—I had been traversing new territory, lands not seen in 1983. Now I was back among the familiar, but, as usual, the familiar had changed. I had planned to stop for lunch at the small café Larry and I patronized in Bruce Mines; the town had been yuppified, and the café was now a boutique-style "General Store." I recognized a side road we had used to get closer to the North Channel waters, and turned down it: the informal parking area in the woods where we had left the car while we wandered the shore was now somebody's house, and there was no place to park whatsoever. I pulled the Escort onto the nonexistent shoulder, leaving its left half sticking out into the traffic lane, and we wandered a bit anyway. Back on the main road we passed several rest areas, and I think most of them were new since 1983; but they all followed the Law of Highway Engineers which says that roads are to get someplace, not to go there, so they rarely allowed us to stop beside scenery. We would pass splendid little Canadian Shield lakes and splashing streams and waterfowl-graced wetlands with nary a place to park; then we would come to blank woods or undistinguished riverbanks, and the parks would be there. Those who site Canadian rest areas seem to be cut from the same unimaginative cloth as their U.S. counterparts.

Thessalon was an improvement. I don't know who named the Thessalon River or the town near its mouth, but whoever it was clearly had Greece in mind. It is not a bad comparison. In the town's harbor, the Adriatic-blue waters of the North Channel lapped little polished islets of light-colored granite. Small boats bobbed and twisted on the swells; boathouses tucked among boulders. A floating bridge danced across water to one of the islets, where two or three picnic tables sat on the bare stone. Small bits of soil here and there on the islet's surface snuggled flowers in their arms. In a few places there were blackened circles of stone eighteen inches to two feet in diameter, with charred bits of what appeared to be curly dock in their centers. Like the Thessalonikans of history, these modern Thessalonikans were fighting off invaders.

At 2 P.M. we came to the city of Sault Ste. Marie, Ontario (population: 80,000), separated from the Michigan city of the same name (population: 15,000) by the broad St. Marys River. The air was clean and hot, and the sky was full of little white clouds. We found our way through quiet Saturday streets to the riverbank. The St. Marys is big in volume, but it is among the shortest streams on the continent. Nearly all of it that can properly be called "river" lies within the city limits of the twin towns built beside its rapids, the long stretch of churning white water the voyageurs named for the Blessed Virgin. Below the rapids the river quickly loses current, blending into the maze of islands and channels forming the upper end of Lake Huron; above, the flat waters of Pointe Aux Pins Bay reach almost to the brink. Pointe Aux Pins Bay opens into Whitefish Bay, and Whitefish Bay—a Lake St. Clair–sized body of water—opens into Lake Superior.

To the voyageurs this great *Sault de Sainte Marie* was a barricade, a wall of white water where paddlers had to portage canoes and goods for nearly a mile over land. To avoid this carry—and to bypass the gauntlet of merchants, hawkers, portaging services, and other businesses that had grown up along the portage trail by the end of the eighteenth century—the North West Company opened a lock on the Canadian side in 1797. It was thirty-eight feet long and could accommodate one freight canoe. American soldiers burned it during the War of 1812. To make sure they got their point across, they burned the house of the lock's builder, John Johnston, as well. That more or less discouraged lockbuilding here for another forty years.

By the middle of the nineteenth century, though, as sail and steam replaced canoes, it became clear that unloading at the bottom of the rapids, passing goods and stores overland, and reloading a second set of ships for service on Lake Superior was not an intelligent approach to naval transportation. Two new locks, the 350-foot-long State Locks, were built, this time on the American side. As shipping and ships both grew, more locks were built, and the State Locks—combined into one and rechristened the Poe—were lengthened, widened, and deepened. Today there are four parallel locks bypassing the Sault (or, as English speakers often spell it, the Soo). The smallest, the MacArthur Lock, is 800 feet long and 80 feet wide; the largest, the four-times-remodeled Poe, is 1,200 feet long and 110 feet wide. Both the MacArthur and the Poe will handle boats with a thirty-foot draft—three feet more than the depth of the shipping channel through Lake Erie.

On the Canadian side, overshadowed by the all the frenzied construction across the river, a single 500-foot lock was built in 1895. In 1983 that lock was still operating, though few freight vessels would fit through it anymore. In 1987, during routine maintenance operations, a large crack was discovered in one wall and the old lock was closed for repairs. It never reopened. In its place—within its walls, actually—Canada opened a new lock in 1998. In defiance of the North American credo which states that all new construction shall be longer, wider, higher, and deeper, Canadian engineers made their new lock shorter, narrower, lower, and shallower. It gives pleasure boats a way to bypass the Sault without fighting 1,000-foot freighters, and it keeps a working lock at the center of what is now Canada's Sault Ste. Marie Canal National Historic Site. Anyone looking for more than this will have to go elsewhere.

Amidst all this lockbuilding activity, the Sault itself has become somewhat lost. It cannot be seen at all from the American side, where four huge locks and their accompanying equipment block all chance of even a glimpse of white water. We tried from Canada, from the visitor center for the National Historic Site, located in the restored nineteenth-century lock administration building. A marked path, the Attikamek Trail, crossed the top of the thick lock gates and traversed South St. Marys Island, among overgrown piles of rubble left over from the original canal construction, to the edge of the narrow channel in front of Whitefish Island. There the path stopped. Whitefish Island, which actually fronts the rapids, is owned by the Ojibwa Nation, and so far they have not authorized the construction of bridges and trails. A little ways upriver, beneath the soaring concrete arches of the International Bridge, there was a view past the tip of the island;

from there we could see the thin white line on the water that marked the upper edge of the rapids. Traffic noises from the bridge blocked any hope of hearing the falling water.

Back at the visitor center we asked the young woman at the information desk if there was any way to see the Sault from the bottom.

"Sort of," she replied. "There's a shopping mall a little way downstream. If you drive around back of the mall, you can see the rapids from the parking lot."

"From the parking lot," I repeated.

"Yes." She smiled apologetically. "It's still not a very good view. Actually, the only place you can see the Sault well is from the International Bridge. If you're going across this afternoon, look to the left."

We found the mall and drove around to the parking lot. There was a riverwalk, and from the walk, as predicted, a view of the bottom of the rapid. It was nearly a half mile away, and from that distance it looked like nothing more than a little splashy water. This was the great Sault, the barricade to commerce on Lake Superior. Expert voyageurs occasionally ran it in light canoes, for sport; they never took a loaded boat through. Fish congregated in huge numbers at its base. Today it is largely ignored. Much of the river's flow now runs through locks and power-plant penstocks; only a fraction goes over the falls. The volume is still far too great to treat with lampricide, which is why the St. Marys River is now the Lakes' biggest lamprey nursery. But it is barely a ghost of the torrent of white standing waves to which the voyageurs gave, in awe, the name of the Mother of God.

We crossed the International Bridge, looking to the left. Opposing traffic was in the way, but we did glimpse, far below, the long maelstrom of the Sault, and it was in fact the best view we were able to obtain all afternoon. On the far side of the bridge the bored border guard barely glanced at us.

"Citizenship?"

"U.S."

"Go ahead."

Along the old portage trail, now Portage Avenue, the descendants of the hawkers who once fleeced the voyageurs were busy with the tourists. We wandered in and out of boutiques and gift shops (looking for film), stopped in a bookstore (looking for research materials), and found a fudge shop (no excuse). The southwestern edge of town had suffered a blizzard of big-box stores and a flurry of fast-food chains:

our motel was tucked in among these, a far cry from the view of Clear Lake we had enjoyed the night before. But there was a copse of beeches outside our window like the one in Rochester, and a field of twilit thistles among which finches were foraging. Just before it became too dark to see, a skunk passed through the thistles and across the adjacent parking lot, the white stripe down its back almost glowing. We have lost much to sprawl, but there are still pockets of beauty present if only we are willing to look.

VII
TO THE SUPERIOR OCEAN

When I began planning for this trip, I had only one fixed rule: that it should conclude with Lake Superior. The order of the others did not matter. The Lake that Longfellow's idealized Ojibwa called *Gitche-Gumee* was going to have to come last.

Anyone who has seen Superior will know why. This is not only the largest body of fresh water on the planet: it is, hands down, the most beautiful. Cold, clear waters crash against cliffs; the eye hunts in vain for a horizon. In the mornings, fog rolls inland, fingering among dark headlands; in the afternoons, whitecaps sparkle on blue water and gulls gambol over long beaches in the winds that blow from the sea. Evenings bring calm, and long cool light. The yellow brick road of Oz lies on the big Lake. The sun hits the water, hisses, and goes out; what was bright gold turns instantly to burnished silver. The sky fills with a bonfire's breath of hovering stars.

Though an arm of Superior reaches almost to the brink of the Sault, the Lake properly begins at Whitefish Point, seventy highway miles to the northwest. Until then one is beside Whitefish Bay, a wide dream of water, but bounded. If you squint you can make out Canada, distant but visible, on the far side. At Whitefish Point, though, the land stops and the sea begins, and there is nothing out there but an imagination of distant shores. Fifteen miles off this point, struggling in huge seas, the ore boat *Edmund Fitzgerald* broke apart and plunged five hundred feet to the bottom in November 1975. It is a measure of Superior's vast size that the *Fitz*'s chroniclers will tell you she almost made haven. In Lake Ontario, fifteen miles out gets you more than a third of the way across. In Superior, fifteen miles out is practically inshore.

In 1983, Rod Badger and I came to Whitefish Point toward the middle of a late July morning. The fog had not yet burned off the Lake; gray surf rolled out of it onto gray sand. Sawgrass moved in the wind. The foghorn at the Whitefish Point Light gronked as if mourning a lost companion. We sat on the solitary beach and played

A laker passing Whitefish Point, Michigan, near the grave of the *Edmund Fitzgerald*.

a tape of Gordon Lightfoot's *Wreck of the Edmund Fitzgerald* and felt as though we had come to the end of all lands; that beyond would be only the sea, and perhaps, far out, the misty byways of the Irish Isles of the Blest.

In 1998, I set out to re-create the experience. I knew the fog would not be there unless the weather cooperated, but I could at least hope for the solitude, and I had the same tape along. What I was not prepared for were the changes that had come to the shore. I had seen the pattern developing on the lower Lakes, of course, but I had managed to cling to the belief that Superior—the largest, loneliest, and most distant of the Lakes—would somehow remain immune. I was wrong.

We left the Soo early, under a sky as clear and blue as my memory of Superior. Twenty miles west, at Brimley, I quickly located the small backwoods motel Rod and I had stayed in fifteen years before. It had a new coat of paint but was otherwise unchanged. So far, so good.

But then came the casino. Two casinos, actually: the Bay Mills Casino in Brimley and the King's Club Casino along the shore two miles out of town, both run by the Chippewa Nation. The King's Club Casino lays claim to being the oldest Indian-run gaming establishment in the United States. I admit to a certain perverse satisfaction in watching Native Americans get something back by exploiting the same sort of moral weakness in Caucasians that the Caucasians often used to divide the natives and appropriate their lands, but I don't like what casinos bring to an area in terms of rapid and largely uncontrolled development. The King's Club crouched like a mother hen in the center of a skitter of suburbs and condos. Homes that looked lifted out of L.A. perched incongruously on the shore of Whitefish Bay. Even this I might have accepted, had it remained confined to the general vicinity of the casino: but it went on, and on, and on, house after house after house. Where there weren't houses there were little driveways, leading off toward the water in the middle of bristles of "No Trespassing" signs; and where there were neither homes nor driveways there were "For Sale" placards. We crossed the border of Hiawatha National Forest and the clutter went on, as inholdings and vacation leases. There was no escape.

"And it's not just the loss of habitat and biodiversity," Melody observed, watching yet another woodful of waterfront dwellings slide by. "It's a pollution issue. All those sewage systems and septic tanks—no matter how good they are, that's an *immense* extra load of nitrogen and phosphorus."

"Not to mention the nonpoint runoff," I concurred.

"Not to mention the runoff. All those roofs. All those driveways. Soil infiltration has gone bonkers."

Occasionally we would cross creeks, and usually if we looked toward the bay at these crossings we would see small wetlands. At least they were protecting those. But wetlands are part of a larger mosaic of habitats which includes woods and meadows and beaches and open water. We use all of these, ourselves: I wonder why we think wildlife can do fine with just one? Protecting wetlands is a start, but they need context. I do not believe in locking development out entirely: Thoreau built a house in his Walden Pond wilderness. But there is such a thing as proportion.

By Paradise, which looked like hell, I had pretty well given up the possibility of getting any solitude at Whitefish Point. There were eleven miles to go. I stuck the Gordon Lightfoot tape in the Escort's cassette deck. Barely two verses into the song we came upon a cluster of cars behind a slow-moving behemoth of an RV, and I lost the train of the lyrics while working my way to the head of the line and around. Grimly, I reeled the tape back and started it over. This time Lightfoot managed to get through three verses before traffic took my mind off him. I jabbed the tape into silence.

At Whitefish Point itself we circled the jam-packed parking lot twice before heading down the highway to the overflow lot. That was nearly full, too: I crammed the Escort into one of only two tiny available spaces. We walked back to the beach, which was noisy with people. Offshore a large laker made its way steadily northward out of the bay into the big Lake, the faint plumes from its twins stacks spouting straight up in the still air. Foot-high waves rolled in like an afterthought. Back near the parking lot we grabbed the only empty picnic table in sight for a quick snack. A pair of sharp-shinned hawks flew over, screaming, and disappeared into the nearby woods. I wondered what they were thinking.

Whitefish Point is one of the great raptor-watching sites of North America. Northward migrants are funneled up to the point as they strive for the narrowest crossing to Canada, here only sixteen miles distant across the mouth of Whitefish Bay. Coming south they repeat the journey in reverse, traveling down the coast of Canada until they see the point across the water and know they can cross over. As many as twenty thousand hawks of various species may pass through here in the course of a single two-month spring migration season. Songbirds follow the same path, and waterfowl, though less restricted by the need for land, also migrate across the point.

As at Point Pelee, the concentration of birds has brought with it a concentration of birdwatchers. There has been a bird observatory here, affiliated with the Michigan Audubon Society, since 1979; lately, bowing to the crowds that now gather at the point, the observatory has begun running a small combination visitor center and gift shop. Knickknacks beside my *Edmund Fitzgerald* beach? I went inside.

"How long has this place been here?" I asked the woman behind the counter. "I don't remember it from when I was through before."

"The gift shop's been here since 1992," she said. "The bird station's been here a little longer." She was a large blond young woman who might have been in her late twenties. Her name tag said "Dorette."

I nodded. "I'm doing a book on the Lakes. Do you mind if I talk to you?"

"Not at all. What kind of book is it, a travel guide?"

"Actually, I'm looking at the changes during the last fifteen years."

Dorette waved a hand at the packed parking lot. "Well," she said, "you'll find a lot of them here."

"So I've noticed. When I came through in 1983, there wasn't any-body else out here."

"It's a little slow now," she smiled, "but it'll pick up later. Most days by midafternoon there's cars parked a quarter mile down the road beyond the overflow lot. I remember when I was a kid, you'd be lucky to see one other car up here."

"Sounds like you grew up close by."

"My grandfather settled here in 1910, and he married my grandma in 1914. We've been here ever since."

"So you've seen a lot of changes. All those new homes along the lakeshore—"

"Right. Retirees come up here and build them, but they don't stay. One winter, and one spring with the mosquitoes, and they're out of here. That's why there's all those 'For Sale' signs. But it seems like there's always someone new to buy when they want to sell."

"They're building awfully close to the water," observed Melody.

She nodded. "And it's getting closer all the time. One guy lost seventy feet off his property last year. The water was practically up to his door."

"Nobody understands geology," I said.

"They don't try to understand. They come up here and they put up the big houses and the big security lights, and then they ask why they can't see the stars. Put up 'No Trespassing' signs and then wonder why everyone's so unfriendly."

"Plant grass so they can mow it," said a woman who had just come up to the counter. "Dorette, do you have any of those bird coloring books left? I want one for my grandson."

Dorette went off to help the woman locate the coloring book. When she came back she said, "Piping plovers don't nest here anymore. You can see why. All the people, and the Jetskis coming by, and everything. There's still some nesting pairs down at Vermilion—I think they're still there. Last I knew, the gate to the beach hadn't been opened yet."

"With all the development coming in, do you ever think of moving somewhere else?" I asked her.

She looked at me. "I had a chance to go to L.A. a couple of years ago," she said. "I got off the plane at the Los Angeles airport and I said, *'What is that smell?'* And someone said, 'It's Los Angeles.' I couldn't get back here fast enough." A wide smile. "I live at the end of an unmarked dirt road in the Hiawatha National Forest. I see deer and bobcats. And coyotes. Coyotes keep me awake. It's real quiet. I can't stand cities."

Outside, more cars were arriving, and there was a hubbub of human voices. Dorette's home may have been quiet, but her workplace was anything but. I wondered, under the circumstances, how long any birds would stick around for the observatory to observe.

So must we place a gate—literal or figurative—across the road to Whitefish Point, and let only *so many* people in? I profoundly hope not; but there may be no other effective way to combat that old nemesis, the Fallacy of Composition. All those retirees Dorette and her coloring-book customer marveled at came here for the same reason the old-timers stay: to be kept awake by coyotes. They will not stop wanting this, nor should they be asked to stop wanting it. Ricki Herdendorf was right: there is nothing wrong with wanting to live in a place like this. But that doesn't mean that all of those who want to can actually do it. The problem is not legal, or distributional, or moral: it is statistical. There simply are not enough coyotes to go around.

By following little roads through the forest, it is possible to remain fairly close to the Lake Superior coastline from Whitefish Point to Grand Marais, the small town that forms the eastern portal to Pictured Rocks National Lakeshore. Melody and I did not go that way. Part of the reason was historical: Rod and I had stayed south, on the pavement, and I wanted to compare what we saw then to what was present today along the same route. The other reason was practical, or perhaps practical tinged with a little hysteria. For the first time since the Thousand Islands we had no overnight reservations. Pictured Rocks has several campgrounds, and I wanted to get there soon enough to have a chance of obtaining a site. Pavement would get us there faster.

So we flew west on state roads: 123 to 28 to 77, Newberry to Seney to Grand Marais. The road passed through a large dune field left over from the Nipissing stage of Lake Superior—stabilized now and covered with good-sized trees—and came out among broad marshes, Nipissing leftovers as well. Newberry, which I remembered as a decaying little shut-up town, had revived itself with a McDonald's and a Pizza Hut and a Subway and an immense Comfort Inn. Seney, where Ernest Hemingway's Nick Adams alighted from the train to set out for the Big Two-Hearted River, seemed little changed in fifteen years—or indeed, since Nick Adams—but Grand Marais had grown. I was not surprised.

Grand Marais occupies one of the loveliest settings on the Great Lakes, a small sand-pond harbor behind a long barrier dune that looks like North Carolina's Outer Banks. Heaving the headlong rush to camp aside, we sought the beach down the paved path that, fifteen years ago, had been a little rutted road with a primitive parking lot at its end. The parking lot, at least, was still primitive. The beach was clean and uncrowded. The view was blue on blue, a huge sky with the Superior Ocean stretching away beneath it, marram grass on the dunes singing in the wind. Westward the tan chiaroscuro of the Grand Sable Banks rose nearly four hundred feet out of the ocean, blurred about the edges by blowing sand; it faded into the distance like a dream of giants. The surf was running only about four to six inches high, but that did little to diminish the experience. It is not possible to face Superior from the beach at Grand Marais and call what you are seeing a "lake." Far too much of the view is water.

The only sour note was sounded by the homes on the dunes. There weren't a great many of them, but I had noticed none in 1983. It was not just the intrusion they represented into an otherwise wild scene of sea and sky; it was geologic ignorance as well. Dunes are like standing waves in a river: their shape is relatively constant, but the substance that forms them is constantly passing through. Houses interrupt this flow, with disastrous consequences. It is only a matter of time before sand fences are sought to keep the sand from piling against the upwind side of the structures, and riprap is required to prevent the sand under the foundations from seeping out. And with these "protections" in place, the dune—a living landform—dies. Why do we still allow these things? Our knowledge of the consequences is not new: the Bible warns against building houses on a foundation of sand.

But there was still a campsite to be obtained, and to get there we would have to face the dubious delights of County Road H-58, a legacy of the uneasy relationship between the National Park Service and Alger County, within which all of Pictured Rocks National Lakeshore lies. When legislation creating the National Lakeshore passed in 1966, it included a provision to build a "scenic route" from one end of the park to the other along the Lake Superior coastline. This, it turned out, was not a popular idea. Environmentalists scorned it as an invasion of heretofore wild lands; engineers scoffed at it as a project somewhere between impractical and impossible to build and maintain. Anti-tax groups chimed in to excoriate the costs. The project was delayed, and delayed, and redesigned, and redesigned. Eventually it was reduced from its original forty-three-mile proposed length to a single thirteen-mile stretch along the rim of Miners Basin, near the west end. When hearings were held on that proposal, 97 percent of the testimony taken opposed constructing even that much.

In the meantime, the park's visitors were using H-58. Alger County had originally built H-58 as a relatively direct means for going from Grand Marais, in the eastern part of the county, to Munising—the county seat—in the west. Like most back roads in the eastern part of the Upper Peninsula, it consisted of gravel laid on a foundation which combined sand from ancient Lake Nipissing with unsorted glacial drift left behind by the continental glacier. It was not meant for heavy traffic, but as word of the park spread it began to get it anyway. By the 1990s, half a million people were visiting Pictured Rocks every year. Nearly all of them came in via H-58, which continued to be maintained by Alger County. As of 1997, according to Park Service testimony before Congress, not one penny of National Park maintenance money had gone into the only through road Pictured Rocks had ever had.

The financially strapped county appealed to the Park Service to take some of the maintenance responsibility. The Park Service refused, on the grounds that it had no authority to maintain roads outside park boundaries. The county suggested that the park could at least take over maintenance of the section *inside* the boundary. The Park Service said they would, but only if the county deeded the road to them: as long as it was county property, it would be the county's responsibility. The argument raged back and forth: the road deteriorated. When Rod and I passed through in 1983, it was already pretty bad. It got worse. By 1997 it was so far gone that the county announced plans to close and abandon it—leaving the majority of the park without any road access at all.

Congressman Bart Stupak, representing Michigan's Upper Peninsula, introduced a bill, H.R. 351, which proposed to remove the offending shoreline road from the park's authorization and replace it with language designating $9 million in federal funds to repair and pave H-58. It had the support of the Wilderness Society, the National Parks and Conservation Association, Taxpayers for Common Sense (the $9 million represented a $4 million saving over the cost of the shoreline route), the U.S. Fish and Wildlife Service, Alger County, and the cities of Munising and Grand Marais. It did not have the support of the Park Service, which still wanted the road deeded to it first. It died without ever attaining a floor vote.

Stupak attached the authorization to a second bill, H.R. 2400, the Transportation Equity Act for the 21st Century. On June 9, 1998, as I was in the middle of preparations for my trip, H.R. 2400 was signed into law and the funds to fix H-58 finally became available. "Available," however, is an odd word to apply to federal funds: it means you can squeeze money out of the Treasury if you can figure out how.

At the Grand Sable Visitor Center, a short distance out of Grand

Marais, we stopped to inquire about campsite availability. The ranger thought there might be several sites left at the two coastal campgrounds, but he was unsure: there had been no recent reports. "We don't make regular checks," he said. "The road is vile."

"I know about the road," I said. "I was here in '83."

"'83," he said. "Well, the road hasn't changed much—or at all."

"Probably hasn't even been bladed," I said.

He grinned. "Close."

A short distance past the visitor center, at Grand Sable Lake, the pavement gave out. The road was squeezed between the small lake on one side and the backside of the Grand Sable Banks on the other. The Banks are a kame, a pile of unsorted glacial rubble which had once filled a large crevasse that formed here at the close of the Ice Age, when the glacier was stagnant and melting back from the south. On top of the Banks spreads a large dune field, the Grand Sable Dunes, extensive enough—and dry enough—that it is sometimes referred to as the "Michigan Sahara." Sand from the dunes blows down the face of the Banks, creating the blurred effect we had seen from Grand Marais. It also blows down the back. The road was covered with it, smoothing out the potholes. For the moment, all was well.

But then we came to the birch/maple woodland beyond. Here the potholes were big and prominent. They were also nearly invisible in the pattern of light and shade the leaves of the canopy threw across them. No speed was slow enough to avoid them. I drove at a steady twenty-five miles an hour. The tires thumped: the car rattled and shook. After what seemed like a couple of years, but in reality couldn't have been more than twenty minutes, we came to the Hurricane River Campground. A lovely loop in the woods whose beach is a spectacular shelving of colorful Jacobsville sandstone, Hurricane was my site of choice. I had camped here with Rod in '83, and again with the family in '87. This year it was full. We agitated back onto H-58.

Seven agonizing miles further—the last three locked behind a slow RV—we came to the turnoff to Twelvemile Beach. I willed the vehicle ahead of us to keep going. It did. We glided down the mile-long side road into the campground with another car on our tail, and I'm sorry to say I drove down the center of the road to keep it from passing. There was one site left. We grabbed it.

As I was getting out of the car to register, the vehicle that had been following us stopped beside us on the campground loop. The woman in the passenger seat rolled down her window. "Are you planning to camp there?" she asked.

I said that we were.

She sighed and rolled up the window, and they drove off. I fought down feelings of guilt. This is the behavior into which we are forced by National Park budgets which do not begin to keep up with National Park demand.

That evening the beach was beautiful, with foot-high combers rolling in from the direction of a blood-red sun appearing and disappearing behind bars of cloud as it sank toward the sea. Pools of water on the beach turned red, too. The transparent waves seemed filled with light; the black clouds were yellow about the edges. Far out over the water a curtain of falling rain played like a cat with the last rays of the day's departing light.

The following morning was damp and gray. The cat had caught up with us, and up close it didn't seem playful at all. We folded the wet tent into the car, getting everything else wet in the process, and headed for Munising. H-58 was a long, linear sea of mud. I crept the car along in first gear, trying to avoid the potholes, which were filled with water. Every few hundred yards, low spots had turned to lakes up to eighteen inches deep. Usually I edged around them, halfway up the roadbank. Sometimes this was impossible: I simply had to aim at what appeared to be the shallowest part and hope for the best. An hour and a half after leaving camp we finally reached pavement and, shortly afterward, the town. Rod and I had found a good nine-point restaurant here, an honest place of fake wood and dark lighting, full of fishermen and locals. I headed for it. The building still held a restaurant, but the name was different and the decor had been upscaled to within an inch of its life. The food had faded somewhat; I gave it a qualified eight. There must be a rule that accounts for that, but I don't know what it is.

There was a spanking-new Tourist Information Center in Munising, half of it occupied by the Park Service, half by the Forest Service. We went in on the Park Service side, and I asked the young woman behind the counter about the current status of improvement plans for H-58. She chuckled as if I had said something amusing. "Money's been approved by Congress," she said. "But there are three levels of government involved—federal, state, and county—and they have to agree on how to spend it. They're trying to decide what sections to pave first, and how wide, and who has responsibility for what. A focus group has been put together. Don't hold your breath."

"That's what they were saying fifteen years ago," I said.

"Fifteen years ago they wanted to build a whole new road closer to the shore, all on Park Service land," she pointed out. "That's been scrapped." Actually, the shoreline road was still technically alive:

it would take yet another rider, attached to yet another piece of only distantly related legislation—this one establishing the Lower East Side Tenement National Historic Site in New York—before Bart Stupak was able to drive the last nail into its coffin.

We drove back into the park, stopping first at Munising Falls. The small visitor center there was still standing but no longer manned: it contained restrooms and display boards. Melody browsed the displays while I walked up the short trail to the falls. The gorge was as I remembered it, fern-hung and full of mist and green smells, but the path had been paved and the loop leading into the sandstone grotto behind the sixty-foot-high falls had been closed off. A large sign explained that the closure was due to liability problems arising from potential rockfall from the grotto's ceiling. Cole Porter may have captured the nation's mood accurately back in the thirties when he wrote "Don't Fence Me In," but things have certainly changed since.

At Miners Beach, surf left over from last night's storm was rolling in as three-foot breakers that pounded furiously aground and hissed up the steep beach to die in the sand. A stiff wind came from the Superior Ocean; gulls were playing in it, soaring directly above the crests of the breakers, facing out to sea. They made continual small adjustments in wing angle and feather loft in order to remain stationary in the air, for no apparent reason but the wind in their faces and the wild challenge of the task. The rain had stopped, and the sun was starting to sneak through the cloud cover. The Pictured Rocks palisaded into the distance like ruined siege walls.

Back at the walls' top we walked a trail along the rim of Miners Basin, over the route of the doomed Pictured Rocks shoreline road, to the place where Miners Falls plunges sixty feet sideways through a slot in the basin's rocky rim. Near the head of the trail, Melody spotted some small yew trees in the understory of the birch/maple woods we were passing through. "I wonder why they were raping the Northwest woods for taxol," she mused, referring to the anti-cancer drug extracted from yew bark that had threatened to decimate yew populations in Oregon before it was successfully synthesized. "There's plenty of yew here."

"Yew think so?" I asked.

She picked up the gauntlet.

"I think they could yewse it."

"But they were looking elsewhere. In the western Yewnited States."

"Maybe there were more yewnits there."

"Did they test it on sheep? On yewes?"

"Yewsually."

"If I stepped behind one of those to take a leak, I'd be yewrinating."

"Well, we *are* in the Yew Pee."

If you've been wondering why I haven't included more of our on-route conversations in this account, now you know. And aren't you glad?

In the late afternoon we drove to Marquette, over hulking headlands and along rocky shores where small beaches lay cupped in concave hollows. The surf had climbed another foot: it was rolling in as trains of four-foot waves, their crests rarely more than ten feet apart. The Lake looked like the teeth of a saw and sounded like an orchestra of kettledrums in which all the drummers had gone mad. In Au Train Bay, two athletic young men were bodysurfing, the sound of their laughter lost in the cacophony of surfthunder. They went under with each wave. We left while they were still in the midst of the maelstrom; I can only assume they made it out alive. Perhaps Cole Porter wasn't so far off after all.

At several spots along Au Train Bay, and again along Marquette Bay, short strips of roadway ran parallel to Michigan 28 on the seaward side. They ended raggedly, at dropoffs: the missing portions of the once continuous roadway had been claimed by the Superior surf. About half the remaining segments had been converted to wayside parks. The other segments led to shoreline homes, mostly new, mostly on the water. At their feet, only yards away, surf chewed away at the bank whose collapse had already claimed much of the roadway. Eyes they have not, neither do they see.

WE LOVE THE BREAKFAST GAME, BUT WE do not always play it while traveling: there are freeway mornings when we are moving too fast, and city mornings when the choices are too broad, and some mornings when we just plain don't feel like going to the trouble. On such mornings we usually end up at McDonald's, which has the twin virtues of ubiquity and consistency. The coffee is pretty good, too.

A thing we have noticed about McDonald's at breakfast time, over the years, is that there are always old men there. The same old men. Affable, irascible, dressed in chinos or jeans, they congregate at one or two tables, swilling coffee and swapping tales, baseball caps perched precariously on the backs of their heads. The caps say "John Deere" or "New York Yankees" or "Not Re-tired, Just Tired." If they were society ladies, one might call them a klatsch. We have never found an adequate term for a klatsch with beer bellies and plumber's cleavage, so over the years we have made up our own. We call this group the Old Laid-back Dudes' Friendship And Restaurant Talking Society. We give the name to you. If it seems unwieldy, you can always get by with just the initials.

We had planned to play the Breakfast Game in Marquette, but we ended up at McDonald's, for reasons which bear examination. I had remembered Marquette as a lovely—if cool—town, a place of art-in-the-park festivals and friendly people. Just a typical little Midwestern city with an ocean outside its door. I was unprepared for what I found this time around, which was a society verging on atomization.

The motel we were staying in provides a good example. It was a member of a large national lower-middle-price chain at which we have stayed in many, many towns. Often our room is near the rear door. This door is usually locked at night. Marquette is the only town we have yet stayed in where it was also locked during the day.

The desk clerk gave us the same pitch, word for word, that she had given to the couple directly before us, and the couple directly before them. All of us were present for all three pitches; once would have done the job. The woman was apparently simply on autopilot, with no attempt to actually connect with the people in front of her.

In the room we found a rotary phone—the only one I have seen in a motel in twenty years—and a small card stating that long-distance calls, even if charged to your own credit card, were subject to a $3.00 minimum fee added to your room bill.

Venturing into town in the early evening, we found two malls. One was down to one or two live stores; the other was only slightly healthier. There was plenty of traffic, but it never seemed to stop anywhere. The usual big-box stores and fast-food outlets on the edge of town were about the only thing in Marquette that seemed to be operating normally—and even that was illusory.

"Look around," I said to Melody over our McDonald's breakfast. "Do you see anything odd here?"

She glanced around the room. "The conversation klatsch is gone," she said.

"Right," I said. "No Old Laid-back Dudes. Where do you suppose they went?"

"They haven't gone anyplace," she pointed out. "They're here. They just aren't talking to each other."

She was right. Four or five Old Laid-back Dudes were actually in the place. They were sitting far apart, staring straight ahead as they sipped coffee, not speaking to one another. In the center of the room a kid of about eighteen sat fidgeting at a table, waiting for an employment interview. The manager busied himself behind the counter, doing his staff's work; the staff stood mostly idle.

"This town is autistic," said Melody. "The clinical definition fits. Difficulty in communicating. Repetitive, meaningless motion. Obsessive acts. It's weird. Was it like this fifteen years ago?"

It was not. Something has shut down here. Economically, what shut down was Sawyers Air Force Base, just outside town: that took place in 1996, but it doesn't fully explain the town's strange state. I've been in towns hit by base and mill closures before. The people don't usually close up as well.

Still tossing these thoughts back and forth, we took the Escort to a drive-through oil change. For the first time in my long history with these things there was no waiting room: we were told, rather brusquely, to stay in the car while it was being worked on. We looked at each other and burst out laughing. Customer service was fine in

Marquette: courtesy and trust seemed to have flown. Where had they gone, and what had pushed them there? We have been able to come up with only one answer: the Huron Mountain Club.

The Huron Mountains, a short distance west of Marquette, are arguably the most scenic section of Michigan. They contain the state's highest mountains. Pictures show gemlike small lakes, polished rock faces, waterfalls, and mile after square mile of stunning old-growth forest. But pictures, alas, are all most of us will ever see. Nearly all of the Huron Mountains are owned by the Huron Mountain Club, and they rarely let anyone else in.

The "club" is a small group of wealthy families, mostly million-aires, mostly from downstate. Henry Ford was one of the founders. Back in the waning days of the Big Cut, they purchased the Huron Mountain tract to keep its forests from falling to the saw. They have run it as a private resort ever since.

And thus Marquette is hammered. Tourism—normally a fail-safe fallback economy as timber and mining dry up and military bases disappear—is hard to do here, because the best of the area's attractions have been sewn up and salted away. "They've finished the resource extraction stage, and they can't go into the next stage," is how Melody described it. No tourism, and only grudging support for local services. A Michigan state law prescribes tax breaks for landowners who allow the public to use their land. Huron Mountain Club members have been trying to figure out a way to take advantage of that law but still restrict public access. Rumors making the rounds in Marquette tell of state resource agency officials and university professors refusing to assist the club in its tax-reduction scheme and being told afterward they are no longer welcome on the club's property to do research. The Huron Mountain Club has saved the scenery of the mountains: that is to their credit. But now it is time to save Marquette.

We got a somewhat different view of Marquette later in the morning when we went to see Bill Robinson. A wildlife biologist—co-author, with a colleague in North Carolina, of the standard textbook in the field—Bill is also a third-generation Marquette native who is passionately in love with his hometown. His reputation is such that he could work anywhere in the world. He has chosen to remain here.

We found Bill seated in his office at Northern Michigan University, surrounded by boxes that were gradually being filled and shelves that were gradually being emptied. A lifetime of teaching was being tenderly sorted out to make room for retirement. "Let's go across the hall," he suggested. "It's not very habitable in here." He led the way into an empty classroom. Stuffed specimens looked down quizzically from the walls; a list of local raptor species was scratched on the chalkboard. We took seats at a study table.

Bill is tall and square-edged and looks as though he was chipped from granite, although after sixty-odd years the granite appears to have undergone a tectonic shift or two. "I've been working with the woodcock," he remarked as we sat down. "When I was just starting out in the field, a long time ago, I met a guy named Bill Marshall from the University of Minnesota. Bill was studying woodcocks, and that was the only thing he wanted to talk about. I thought, 'I'd better not get into those until late in my career. They sound addictive.' Now it's late in my career, and I can afford to be addicted." He chuckled. "Let's see, where were we? I sat down with your book last night and took some notes. *The Late, Great Lakes.* Is this one going to be called *The Later, Great Lakes?*"

I grinned. "Could be. What's in the notes?"

"They're not in any particular order." He consulted a list in his hand. "A lot of these you've probably heard about. The invasion of the zebra mussels, particularly."

"Yeah," I said. "I thought we might talk about that in some depth."

"I don't have a lot of depth on it myself," he protested. "I get most of my information through term papers." The granite shaped itself into a smile. "I had a student in my class who lived down on Lake St. Clair, and she said, 'You've never seen them?' And I said, no, I'd never seen them. So her mother went down to the beach and grabbed some shells and plunked them in an envelope and sent them up. I had an envelope full of them for a while."

"Do you know of any signs of them in Superior?" I asked.

"I haven't heard of any," he said. "But I'm not really in the loop, particularly, so I don't know. The river ruffe has been a problem in the Duluth area—I haven't heard much about it around here." He glanced at his list again. "Sea lamprey are apparently under control, but it's still a very expensive program. They keep improving their methods, but it's not a permanent solution. They have to keep treating the streams with lampricide." A chuckle. "I like that term, 'treating' the stream. Give 'em a piece of candy, or something. It's actually zapping them with poison. Anyhow, that problem has been reduced, but it's been very expensive, and the lampreys are still there. Whitefish in the Lake seem to be doing OK—there's a significant licensed commercial fishery. Lake trout have come back in quite large numbers, and there are some large fish. Lake herring seem to be staging a comeback, too—nobody knows why. One thing that seems to have declined are the smelt. I'm not sure why. But that's an exotic anyway."

"Yeah," I said, "it's not a native fish. That may be a sign that the ecosystem is actually recovering."

"It could be," agreed Bill. "It could be. Yeah. So things look a little brighter, maybe, than they did ten or fifteen years ago." He looked up. "I don't know if you've taken a ride around town or not. The erosion at Picnic Rocks is extreme. I wouldn't mind driving down there, if that would be OK."

"We'd love to," I said.

"OK. Well, I'll show you, and we'll make some comparisons. Commercial shipping, I think, has been the problem. There's a demand for keeping water levels high so ships can have deep water in the harbors. You may recall we were doing some studies back in the eighties on winter navigation at Sault Ste. Marie, and its effects on mammal movements. I haven't been down there for a long time. I suspect the banks are still eroding and being undercut there, from the ships pushing the ice into them."

"Is winter navigation actually going on?" I asked. "I thought that was one thing even the shipping industry didn't really want."

"Well, I think they close the river for about five or six weeks in January and February, or something. We keep getting a thing from the Corps of Engineers to comment on, and the closing and opening dates keep getting closer and closer together." Another chuckle. "I was down at the Soo several years ago with a graduate student of mine named Ron Jensen—Ron was doing his thesis on the effects of winter shipping on birds. There'd just been about a four-foot snowfall. We drove into one guy's place on the shore of the St. Marys River—he was shoveling the snow off the roof of his garage, and we talked to him from the ground as he was shoveling. And Ron asked, 'You ever see any eagles around here?'

"The guy says, 'No, I don't think so.' And then—'No, wait a minute. Sometimes I see those things goin' down, and they nest on an island over there. There's a whole bunch of 'em in nests there. Is those them long-legged jiggers?' "

Bill smiled quietly. "We said, 'No, those are probably great blue herons.'

"And then he asked, 'What do you want to know for?'

" 'Well, we're doing some studies on the impact of winter shipping on birds.'

"'Winter shipping,' he says. 'What a horseshit deal that is. They come by here with the icebreakers and they push the ice onto the shore. They wreck my dock every year and I have to get a permit to build a new one. Pay twenty-five bucks every year to be allowed to build my dock again that they've wrecked.'" Bill chuckled once more. "He wasn't a great—fan—of winter shipping. And I think that's not atypical. The icebreakers do push the ice around like that, so the locals are not real fond of them." He consulted his list again. "The Little Presque Isle battle sort of came and went during your absence. Have you heard of it?"

"No, not really," I said. "I think you took Rod and me out to Little Presque Isle."

"Yeah. Well, the original battle back in the seventies was that they were going to build a power plant there. And on that one, we finally got them to expand the plant here in town. But we have, in Michigan, a nice law called the Land Trust Fund Act. Down in the northern Lower Peninsula there's a lot of oil under state land, and a citizen referendum approved a bill that a certain percentage of the royalties that the state gets from that oil should be devoted to recreational development. Unfortunately, the term 'recreation' wasn't"—he chuckled again—"very well defined. And one of our local state foresters—he's a decent guy, he works for the Department of Natural Resources—got a grant from the Land Trust Fund to develop a campground for motor homes on the point at Little Presque Isle. He brought it to the Natural Resources Commission in May, I think, of 1992, at a meeting downstate, and there was very little public input. And they approved it.

"The Natural Resources Commission comes to the U.P. once a year, and that was their next meeting. They met in Escanaba, and the meeting started at 7:00 in the evening and didn't get out till 2:00 A.M. There were something like sixty people that spoke to the commission, and only about two of 'em spoke in favor of the campground. And to their credit, the commission backed off. They appointed a local committee, and they included several environmental people. I was on the committee. And we met for about two years.

"Now, this guy gets a $300,000 grant to develop that property, and you've got to let him save face a little bit. So the committee recommended that there be no campsites on the Lake Superior side of the property, east of Big Bay Road. But as a substitute we agreed to let him build cabins on Harlow Lake, a mile inland from Superior, and in a few places in the woods around there. They planted some dune grass on the point at Little Presque Isle, which helps stabilize it—put a few rails around, so people don't walk off the edge and trample it to death—it's decent. It's not obtrusive. I think it turned out pretty well, actually. The only minor problem is that people insist on feeding the deer, so they bring out big bales of hay and put them in the woods there by Harlow Lake. We've got the same problem here in Marquette, at Presque Isle Park. People are feeding the deer, and we've got too many of 'em. I'm on the Presque Isle Citizen's Advisory Committee, and we've approved live-trapping and removing to try to bring the deer numbers down to twelve or thirteen, which would be the normal carrying capacity for the half square mile of the park. But so far they haven't caught enough in the spring. And several of the deer that were transported twenty or thirty miles away came back."

"Have you considered a controlled hunt?" I asked.

"Yeah, we've considered it," he said, "but I don't think it'll go. A lot of people just love those deer—it's their life. They don't want to see those deer touched. They've got 'em all named, there's Puffy Face, and Wiggly, and—"

"It's the Bambi complex," I observed.

"Yeah," he said. "Yeah. It's hard. But there would still be—you know—a dozen or so deer, which would be reasonable." For the first time in a while he consulted his list. "Let's see—loons, and lead poisoning from sinkers. There've been some recent discoveries that loons seem to be attracted to lead, for some reason. I saw a film at the North American Wildlife Conference in Washington several years ago when they were first discovering this. They had a loon in a tank, and they had big pebbles on the bottom. And they would put a lead sinker in there, and the loon would go over and play with it, and pick it up—carry it around—I guess they're attracted to it."

"Why would that be?" I asked.

"I don't know," said Bill. "I don't understand it."

Melody looked thoughtful. "It tastes sweet," she said. "Where did I pick that up? I don't know."

"I think most of the loon deaths are probably a result of fishermen losing their fish, and the sinkers are there, on the line or something," said Bill. "The birds die of lead poisoning. I was skeptical at first, but apparently it is a problem. Loon numbers have been fairly stable in this state, and they still are—I think we're talking around three hundred pairs. But we're looking at a goal of at least five hundred pairs for a period of a number of years to remove them from the threatened list.

"Uh—the waterfront. Marquette has been, I'd say, relatively progressive on this, but there are always threats. The park that we have downtown used to be a coal-unloading facility, and we still have coal being unloaded in two places. The citizens—many of the citizens—are opposing what apparently are going to be high-rise twelve-story residential buildings across from Shoreline Park. I'll show you where that is down there. It almost looks like a done deal."

"I was going to ask you about development around here," I said.

"Yeah," said Bill. "As we go along it will remind me of things. You want to take a ride?"

We spent the next several hours in Bill's dark green Subaru station wagon, riding around the Marquette waterfront. The surf was still up from the storm that had soaked us at Pictured Rocks: breakers four feet high rolled in from the blue horizon and smashed themselves into spray on sand and stone and breakwater. Boats in the marinas bobbed up and down and bumped against each other. The sun shone.

We began at the east end of town, at the municipally owned electric plant, where Bill pointed out the effluent canal. "This is a fairly popular bathing beach," he told us, "because the water comes out warm. But right now the Lake itself's pretty warm . . . people have said. I've gotta go and totally immerse myself, in what used to be my annual ritual, although I've chickened out recently." He chuckled. "But it's about time." He gestured at the plant. "So that's pretty ugly, anyhow, and they bring in—well, let's take a drive down to the loading dock."

Down on the dock there was a big pile of coal. The yellow wall of the power plant rose on the right; a large ore dock in a state of serious disrepair loomed to the left. At the lip of the loading dock two men in old clothes, fishermen obviously, rummaged beneath the canopy of an ancient pickup. "They bring the ships right in and unload them on the docks there," Bill said, pointing. "This past winter I had my environmental science class down here. We parked right down there where that pickup is, and we stayed around five or ten minutes, and I discussed it, and then I said, 'OK, let's go.' And a couple of the guys went running over to the coal pile. I said, 'What are you doing?' 'Oh, I want to see what coal looks like, I've never seen coal.' " He chuckled. "I said, 'Yeah, that's a good idea. You've never seen coal? There it is.' They brought back a few lumps."

We drove north along Lakeshore Boulevard to the Picnic Rocks. "This is the place that's scheduled to be developed fairly shortly for twelve-story buildings, which most of the neighbors are not real crazy about," Bill said, pulling the Subaru into a tree-shaded parking lot. We walked across the grass toward the beach. "They're tearing down the old buildings from the foundry," he continued, pointing to a decaying building across the street. "They made marine winches there, and other heavy equipment. I had one of my friends call me the other day and ask me if I wanted the whistle from the plant. I said, 'Why is that?' 'Well, I heard that your grandfather made that whistle.' I hadn't heard that, maybe he had. He was a machinist, and he worked there quite a few years. My father also worked there for a while in the 1940s. He had a music degree from Northern, but he worked at the plant here, as a laborer. But this is where the high-rise will go in—what they call high-rise in northern Michigan. Ten or twelve stories is pretty high for us here. Most of us feel it's out of character, but I think the battle's probably been lost."

The beach was edged by a ragged three-foot scarp of sandy earth. The Picnic Rocks loomed several hundred yards offshore, bits of glacial-polished granite rising like petrified sea monsters from the Lake Superior waves. "When I was a kid," said Bill, "I could stand on the shore and throw a rock and hit the Picnic Rocks. And it's not that I had that good of an arm. It's that—it was like from here to that tree, maybe. Sometimes at low water in the spring you could actually walk across to the first rock. That's all eroded in my lifetime. It's a combination of high water levels and the way the shore's formed. We've lost 100 to 150 feet of shoreline along here."

"It comes and goes," Melody observed.

"Well, it hasn't been coming much," said Bill. "It's been going."

Further north we stopped at a harbor overlook. Waves smashed against the breakwater: in the calm water behind it a ship was pulled up to Marquette's last operating ore dock. The large crane on the ship's deck identified it as a self-unloader, a type of vessel native to the Great Lakes. Bill pointed to the vessel and the sand flat behind it.

"What's happening," he said, "is the sand that we've lost at Picnic Rocks is being deposited around here. We have a very wide, growing beach up here. We don't get the waves coming at an angle this way, because of the breakwall. So this is where most of the sand is ending up, in a place where they don't particularly want it, out where the ships come in. Actually they've exacerbated the process more, because they're dredging for the ships, and they put the sand out in deep water. And of course it never gets back onshore."

The drive ended in Presque Isle Park, a large outcrop of Precambrian granite topped with Jacobsville sandstone, plopped in the

Superior surf like a gigantic cow patty and decorated with dark green forest. A sandy isthmus connects it to the mainland. The low ground of the isthmus sprouts playfields, and a water slide, and the tree-shaded picnic lawns that, in 1983, were full of artists and easels for the Art on the Rocks Festival held each year on the last weekend of July. The rocky hump of the isle, roughly half a square mile in extent, is maintained as a natural area. Forest covers most of it; trails and a narrow paved road wind through the trees. The road is closed to motor vehicles until 1 P.M. each day. We parked near the barricade and walked up into the soft green light of the woods. The hands of maples moved in the breeze. Sunspots played tag with each other across fallen leaves and tufts of short, tough grass. A doe browsed casually near a bright yellow sign that said "Keep Safe Distance From Deer."

"This is sugar maple forest," Bill pointed out. "This grass, apparently, is fairly resistant to browsing. But in a normal sugar maple stand, the ground would be covered with sugar maple seedlings, and a lot of other herbaceous plants. There just isn't much of that here. It's what we pay for keeping the deer people happy, I guess."

We left the road and took a small trail through the woods, headed for Black Rock Point at the park's outermost edge. The trail skirted the tops of tall cliffs with their faces turned to the sea. Ancient cedars, shaped by the winds, clung precariously to small ledges as if posing for an Ando Hiroshige landscape. "Not long ago, there was a young woman who rode her bicycle off the cliff just about here," Bill said quietly. "She died. I'm on the Presque Isle Advisory Commission, and we got calls. 'You've got to put up more signs around there.' But it's sort of common sense, you know? What would the signs say, 'Don't ride your bike over the cliff?' I don't want to downplay the tragedy of that girl's death, but a hundred-foot vertical cliff seems obvious. You can put signs up all over the place, and people will still ignore them. So the city so far has just continued to take its chances."

Superior was throwing itself at Black Rock Point like a frustrated wrestler, flinging spray thirty feet and more into the air. The dark, gnarled granite did not give. There is a school of thought which suggests that these Precambrian granites in Presque Isle Park are the oldest surface rocks in North America, predating even the Vishnu Schist at the bottom of the Grand Canyon. Knobby, weathered, beat about by the sea, they certainly look the part. This was a different Marquette than the autistic city we had been immersed in since our arrival the night before.

Surf at Black Rock Point, Presque Isle Park, Marquette, Michigan

We bought sandwiches and drinks at a small deli and drove out Big Bay Road past the smooth granite hump of the Sugarloaf to Little Presque Isle. Sitting on a log in the sand with the surf surging nearly to our feet, we ate the sandwiches and watched a sanderling and a semipalmated plover work the edge of the rift a few yards down the beach. Waves crashed in the breach between Little Presque Isle and the mainland, cut through during an immense storm sometime in the 1930s: before then the island had been a peninsula. Geology is alive.

Halfway back to the car, Bill stopped under a tree and began making little "pishh" sounds with his lips. Soon the canopy over our heads was alive with small feathered bodies. Warblers. Chickadees. A vireo. A downy woodpecker. Tiny heads cocked, bright pupils seeking the source of the sound, they fluttered closer.

"Impressive," I told him.

He looked sheepish. "Some birders don't think it's fair," he said.

He dropped us back at our car. We drove back, just the two of us, and took the wooden stairway up Sugarloaf in the yellow light of early evening. The leaves of maples glowed like large green lamps, the bulb of the sun sinking slowly behind them. The top of Sugarloaf had been transformed, with rails and viewing platforms where I remembered bare granite. My view of Marquette had been transformed, too. There is hope for the place. Like the rocks of Black Rock Point, Marquette's problems are old and hard and dark. But Superior never gives up. Marquette must not give up, either.

In the midst of the spread-out town, Northern Michigan University's domed stadium—officially the Superior Dome, unofficially the YooperDome, after the once-pejorative term Upper Peninsula residents have adopted as their own—spread its wide roof like a revival tent. Half a mile distant and three hundred feet below us, the surf of Superior flashed white in the descending sun, its rock-hewn rumble carrying clearly to us beneath the banter of small birds and the wild, constant, whipped-cur moan of the mountain wind. Far to the north across the mottled plain of broad water the sky kneeled down and kissed the wide blue beckoning convex curve of the edge of all known lands.

COASTING. SUPERIOR DAYS MELD TO-
gether like melted chocolate, as sweet and warm as memories of
childhood summers. We were four days getting from Marquette to
Duluth, on small roads next to the sea. There were beaches and islands
and headlands; there were roads that ran beneath cool canopies of
white birches. There was the constant companionship of the Superior
Ocean, big and blue and white around the edges, as restless in its
bed as an insomniac squirrel. The days and the miles ticked by; city
and country, cloud and sunshine, night and day. The scene changed
constantly. The Lake was always there.

On one of the days we drove out the Keweenaw Peninsula, a long,
narrow mountain range curved like a scimitar and surrounded on
three sides by water—the Bruce, but made of basalt and conglom-
erate, not dolomite. At its tip, eighty miles from shore, the town
of Copper Harbor clung like a wilderness outpost to the shores
of its rocky bay. Just west of town the Brockway Mountain Drive
edged a backbone of scantily clad stone into roaring winds more
than seven hundred feet above the Lake. Superior makes its own
weather, and the Keweenaw sticks up in the middle of it: more than
seventeen feet of snow fall here during an average year. Summers
compensate by being sunny and mild, with seventy-five degrees
considered a hot day. Fall comes early. By mid-August it was already
making its presence felt—a scarlet splashing of maples, a yellowing
of birches. On the outer curve of the scimitar, facing Canada across
more than a hundred miles of open water, we strolled a pebbled beach
where I had once found fossils and found that day only many other
people. Teenagers leaped waves or broiled themselves on towels.
Smoke wafted skyward from beachside barbecues. At the south end
of the strand a new home thrust itself into the water on a tiny flat
peninsula, precariously protected by enough armor stone to rebuild
Jericho.

The Porcupine Mountains silhouetted in the sunset above Lake Superior surf, Porcupine Mountains State Park, Michigan

On another day we hiked to Mirror Lake in the Porcupine Mountains, the Midwest's biggest and ruggedest wilderness area. The continental glacier ground right over the top of the Porkies, smoothing their summits but taking big ragged bites out of their sides; the holes left by the bites hold lakes and marshes and little cascading rivers and the biggest remaining uncut forest in Michigan. The six-mile round trip seemed short. I tried Bill Robinson's bird-calling trick under a canopy of Porkies maples, hyperventilated, and nearly passed out. Only a few curious birds straggled in. That evening we sprawled on the shore of a small resort near Ontonagon, fiddle and pennywhistle in hand, playing tunes with a bluegrass musician and his new wife up from Madison on their honeymoon. The Superior surf played percussion. A fire leaped and crackled on the sand, and a small, fresh wind blew in from the west, out of the silver sea.

Two days later we walked the shore of Madeline Island, on the edge of the Apostle Islands National Lakeshore, in the raw, wet dimness

Mirror Lake, in the Porcupine Mountains

of a dawn with rain. The sand of the barrier beach at Big Bay, on the island's eastern end, was damp with the rain that had fallen during the night; the surf slapped it absently, as though whacking at mosquitoes. The water of the cedar swamp behind the beach was dark and still, soundless except for the creak of the boardwalk beneath our feet and the almost-rhythmic plunk-*Plunk!* of the drips, now distant, now near, that fell from the soggy elbows of the trees. Close at hand, red sandstone ledges cantilevered over wide, dark, mysterious lairs; the sea surged under them, in and out, in and out, like a tireless lover. A breath of damp mist hung over the island, a harbinger of further rain.

The Big Cut swept through this western Superior country a century ago, and mining has savaged it, but it remains remarkably unscathed. There is a single AOC, Torch Lake, near the town of Hancock at the bottom of the Keweenaw Peninsula: when I saw the lake fifteen years ago it was the color of cinders and its shores were piles of copper slag. This time around, small boys were swimming off a new dock in nearly clear water and the shore was green with young grass scattered with picnic tables. Behind the park, overlooking the lake, homes were going up.

Homes were going up everywhere else, too, of course. Development is by far the biggest current problem on the Great Lakes, and the western section of Superior has not escaped. At all places where the shore was not legally protected there were houses on it. Homes rimmed the Keweenaw at Lake level; homes spread along the bay near L'Anse and ran in a nearly unbroken line from Ashland, Wisconsin, out to the tip of the Bayfield Peninsula. We looked for a view toward the heart of the Apostles from the long north shore of Madeline Island and saw only houses and driveways to houses. The road stayed so far inland the Lake could not be seen. If a view of the Apostles exists from Madeline, there is no public access to it.

New homes do not exist in isolation; as Kent Fuller and Bob Beltran suggested back in Chicago, they bring their support structures with them. These were strongly in evidence as well. There was much more pavement than there had been in 1983, and more fences and railed overlooks protecting the scenery. There were big-box stores and fast-food outlets. A golf course had been cleared in the hilly forest near the tip of the Keweenaw. We didn't see the new sewage facilities and new generating plants and new transmission lines, but they had to be there somewhere. We are not talking about cabins with kerosene lights and one-holers out in the woods; we are talking about three bathrooms, garage-door openers, and Internet access. There is no reason to deny these things to people who live in upper Michigan and Wisconsin, but they do not come without cost.

Some of the changes were positive. When I passed through with Rod, Calumet was a gracile ghost town of handsome but decaying sandstone buildings amid defunct copper mines halfway to nowhere out the Keweenaw Peninsula. This year the buildings were spruce and shiny and the streets were bustling; the downtown was full of shops that seemed to be competing well with the Pamida discount store and Burger King that had come to the outskirts. Other changes were less encouraging. In Wakefield, a good Breakfast Game restaurant had died; the corpse bore a large "For Sale" sign. In Ontonagon, the overworked, trashed city park looked like some I had seen in Chicago. There were traffic jams in Ironwood. The Summit Peak Trail in the Porcupine Mountains had turned into the Summit Peak Stairway, with solid wooden steps and walkways suspending visitors six inches or more above the living rock of the mountain for the entire half mile from the parking lot to the top. Given the use it was receiving, visitor control at that level was probably necessary; but it was one more slice in the death of a thousand cuts the north country has been hit with over the last fifteen years.

The most striking differences were at Ashland. I had remembered Ashland as a sleepy little college town near the south tip of Wisconsin's big, island-dotted Chequamegon Bay. It was still a college town, but "sleepy" and "little" were no longer good descriptors. The outskirts were a tumbled mass of freeway culture—a Wal-Mart, two big supermarkets, a giant discount pharmacy, too many fast-food outlets to count. Suburbs had sprung up. There was much construction on the campus at Northland College. Our daughter Jenny attended Northland for two years in the late eighties: I'm not sure she would recognize it today. At the information desk in the Sigurd Olson Institute, an environmental study center at Northland, I asked the girl behind the desk if the college had changed a lot since 1988. She blinked. "It's changed a lot since last April," she replied.

West of Ashland a short distance there leaped from a wood beside a wetland an immense hip-roofed wooden building with a cupola on top, an unlikely architecture resembling the seed of some wild mating between a California executive's home and a lighthouse. This turned out to be the Northern Great Lakes Visitor Center, a joint project of the U.S. Forest Service, the National Park Service, the State Historical Society of Wisconsin, the University of Wisconsin Extension Service, and several private nonprofit and for-profit entities. The place had been open only since May, and road signs pointing to it hadn't gone up yet. "We have signs ordered through the Park Service," the attendant at the information desk in the bright, barely peopled lobby explained. He sounded apologetic. "We were going to have the state of Wisconsin do them for us, but they took forever to arrive, and then when they did they were too small. We had to send them back." The information desk was a large, gleaming, beautifully crafted piece of cabinetry with a plaque identifying it as being built with lumber cut from logs salvaged from the bottom of Lake Superior. The Superior Water-logged Lumber Company in Ashland has been doing this for approximately six years. The floors of many Great Lakes harbors—including Chequamegon Bay, by Ashland—are jackstrawed with logs that sank during timber-rafting operations one hundred years and more ago. Worth little then, they are worth thousands of dollars apiece today. The fine-grained lumber—ten times the number of growth rings per inch of any timber grown in Michigan and Wisconsin since—has been perfectly preserved by the cold waters of the Superior Ocean.

Water-logged Lumber founder Scott Mitchen began lifting logs off the bottom of the Lake with truck inner tubes in 1992. Today he uses flotation bags and robot-controlled cranes, but the principle is the same. Out of the muck and into the air—and into fine furniture, cabinets, and musical instruments. All without a single live tree being touched. It is not a sustainable supply, of course: Mitchen is mining his old logs, not growing them. While it lasts, however, it will take the pressure off the living forest. And it is likely to last quite a while. It has been estimated that as much as 10 percent of the Big Cut went to the bottom. Almost all of that timber is still sound and sitting down there. Waiting.

In the middle of the afternoon of the fourth day out from Marquette we crossed the big bridge over the St. Louis River at Superior, Wisconsin, and pulled into Duluth, Minnesota, the westernmost city on the Great Lakes. It had been a largely gloomy day. There was rain: sometimes light, sometimes heavy, always there. The Lake was a restless gray plane fading into mists. Beautiful but moody. We had left her mostly alone.

This day had begun with a ride on the tiny, tossing ferry from Madeline Island to the mainland across the wide, slate-colored surface of Chequamegon Bay. Bayfield rose to greet us, huddled on its hillside in the moist, cool air. The Apostles were smudges scattered across the northern horizon, fading slowly in and out of the thick atmosphere like cinema images trying to become substantial. The diesel ferry huffed and stunk. Ashore, we drove to the Sand Bay Visitor Center in the Apostle Islands National Lakeshore and walked around a bit in the rain. A small set of barrier dunes in front of a big wetland. Sea kayakers out in force. A general mistiness. The visitor center had recently undergone a large-scale renovation—so recent that the restroom doors had no signs on them yet. The old, outdated building they were renovating had barely been on the drawing board in 1983. Time is funny.

We stopped briefly at a little picnic area at the west end of park, where a whole bevy of kayakers was gathering for a run to the popular Squaw Bay sea caves. The kayaks lay on the beach like long colored cucumbers, their prows on the sand, their sterns levering up and down in the swells. The kayakers clotted inside the single picnic shelter. Rain fell, light but insistent. One girl stripped down to a

Chequamegon Bay from the city park at La Pointe, Madeline Island, Wisconsin

purple swimsuit, thought better of it, and put her clothes back on again. On to Cornucopia, "Wisconsin's northernmost village," where we ate lunch under a picnic shelter beside a flowing artesian well and watched loons play hide-and-seek off the soggy beach. Raining in earnest, now. We switched drivers and Melody drove into Duluth, over the up-and-down rolls of Highway 13, the big Lake barely guessed at behind a damp barrier of woods. We did the chores—laundry, dinner, journal and camera maintenance—and turned in. Rain pelted the motel roof and hissed at the tires of cars passing on the road between our harborfront motel and the harbor. Tomorrow the North Shore; then Isle Royale.

I am acutely aware of the approaching end of the trip. Each view of the Lake has poignance, as I attempt to pull as much of the sight and sound and feeling of this body of water into my pores as possible. The stops lengthen, and the moments tick by, and I absorb and absorb, and am still not satisfied.

We left Duluth early, under clearing skies. The rain that had followed us west from the Apostles and kept us company all night had capered off to play elsewhere, and the clean-washed air carried a hint of conifers. Just above Two Harbors we stopped to stroll the dark-pebbled beach at Flood Bay State Wayside and watch the last luminous clouds shift and dissolve around the dark bulk of the great headland known as Silver Cliff. Gulls swooped and screamed; pebbles shifted in the surf, making a sound like poured roofing nails.

As with Ashland, Two Harbors was a shock. I thought I knew the place: I had spent three days there as recently as 1994. A small, sweet town, a bit old-fashioned, tucked neatly around its two namesake bays, its only concession to freeway culture a Subway outlet on the Highway 61 bypass. Change can move remarkably fast, however, and four years is enough for a major overhaul. The bypass had sprouted a full-fledged forest of fast food: McDonald's, Burger King, Taco Bell, Dairy Queen. There was a big new discount store and an even bigger supermarket. Downtown, the Ben Franklin was holding a liquidation sale. Many stores stood silent and empty; many more had shifted merchandise and clientele, and had become boutiques. Raw economic data suggest that the town is doing well. Since 1983, when Rod and I passed through, the unemployment rate has dropped from more than 17 percent to under 7 percent. But raw economic data doesn't tell you what happens to a town's heart when it switches from staples to souvenirs, and it doesn't measure the intangible but troubling difference between minimum-wage jobs at Ma and Pa's grocery and minimum-wage jobs at Megafood. There is no question that Two Harbors is economically healthier than it was fifteen years ago. There is some question as to whether it is still Two Harbors.

At Gooseberry Falls there was a new visitor center, a realigned highway, a much larger parking lot, and some talk of turning the old visitor center—now closed—into a trail shelter. Two evening grosbeaks, both female, cadged sunflower seeds from feeders beside the trail to the falls. At Castle Danger, a little restaurant where Melody and I had breakfasted in 1994—a good, strong 8.5—had ballooned outward and become a vaguely Scandinavian log-cabin-style Big Restaurant, dwarfing the original structure, which remained fastened to its side like a squat, confused lamprey. All the way up the coast there were subdivisions, some new, some newer, some so new they were only signboards and dreams. While I drove, Melody wrote down the names. Spinnaker Inn. Breezy Point Resort. Future Site of Grand Superior Lodge—Resort Homes For Sale. Superior Shores. Cove Point Lodge. Superior Ridge. The Cliffdweller. Surfside. Aspenwood. Bluefin Bay. We seemed to have come a long way, indeed, from Cozy Food, Warm Beer, and Cold Rooms in Massey, Ontario.

But always there were the bluffs and the big Lake bordering them, the Superior Ocean, sweeping off toward unknown lands to the east. Rocky headlands rose from the surf; lighthouses crowned promontories. Seabirds tossed past. Green walls of mountains rose to the west, stepping ahead of us upshore into the misty distance. Rivers rumbled and roared through green gorges and ran out past stony beaches to lose themselves in the immense cold loneliness of the sea.

Superior is one of the world's great seascapes, and Minnesota 61, which borders it, is among the world's most scenic coastal roads, comparable in the United States only to U.S. 101 along the Pacific or Route 3 around the outer rim of Maine's Mount Desert Island. Rising and falling and twisting and tunneling to stay as close to the shore as possible, the two thin lanes of pavement bore through forests and loop around small hidden coves. Overlooks abound, well placed and often protected by stone guardrails, usually with good access to the water. There are twelve state parks and waysides in the 150 miles from Duluth to Canada, ranging from tiny Flood Bay (20 acres) to vast Tettegouche (9,000 acres), whose big landscape embraces beaches, headlands, wave-cut arches and caves, rivers, peaks, and several mountain lakes.

I have used the word "mountain" often with regard to this shore. The word may seem presumptuous when applied to a set of peaks whose highest point barely tops twenty-three hundred feet. It is also technically incorrect: the high land along the Minnesota coastline is not a cordillera but a cuesta, the upturned edge of the plateau that forms the flat thousand-mile-wide heart of North America and which is known, in its treeless western section, as the Great Plains. But technicalities are one thing, experience another. I have spent much time with the Oregon Coast Range. The Minnesota coast range is eerily similar, right down to the basalt both are primarily formed of. The Oregon basalts are geological babies, however—barely 4 million years old—while these Minnesota basalts are ancient. They are the remnants of massive Precambrian sheet flows, and they came out of the earth's womb more than a billion years ago. Amazingly, this old rock still shows the signs of its formation: hexagonal cooling cracks, columnar jointing, and the frothy vesicles that formed near the top as the flow expanded rapidly on contact with an oxygen-deficient atmosphere no living thing had yet begun to breathe.

We drove up Palisade Head, where a 350-foot cliff made of billion-year-old columns drops straight to the white surf and you stare down on the gray backs of gulls soaring high above the sea but still beneath you.

We drove around the backside of Tettegouche Park to Lax Lake, the only one of Tettegouche's small inland lakes that is accessible by car. Trails led off to Mic Mac Lake, and Tettegouche Lake, and Raven Rock, and Palisade Valley. We passed. It was a hot day, and none of them was wild Superior. Perhaps some fall, in color season, we can come back.

We wandered the beach at the mouth of the Temperance River, named for its resistance to developing a bar. Some sources, in fact, claim that this stream "never" has a bar across its mouth: an obvious falsehood, for we saw one there. Perhaps the river has finally received word that Prohibition is over. A busload of Japanese tourists stopped at the same time we did, and a gaggle of small boys spurted from the bus to the shore and—yes!—began throwing rocks in the Lake. I thought of the inner-city kids back at Lakeport State Park, on Huron. No matter what the atheists may think, there *are* universal truths in the world.

Toward midafternoon we arrived at Grand Marais, the northernmost U.S. town of any size on the Great Lakes. Small, old, and pleasantly unkempt—I mean the word as a compliment—Grand Marais

Shovel Point from Palisade Head, on Minnesota's North Shore

droops casually around the shore of one of the most attractive small-boat anchorages in the country. Canoe liveries and outdoor-goods stores give it the feel of a North Woods outpost. Mixed conifers and hardwoods crowd against the town and climb the high broken-off edge of the Great Plains, labeled the Cabotian Mountains on early maps, known today as the Sawtooth Range. Sprawl has begun to knock at Grand Marais's door, but so far the town hasn't answered. The subdivisions carved in the woods are still overwhelmed by the woods themselves. The only concessions to freeway culture we could find were a small Subway outlet and a Dairy Queen that closed for the night at 7 P.M.

Our motel lay three miles out of town to the northeast, in the woods, a small, square, yellow building with eight slightly tired but scrumptiously cozy units. A smidgeon of Lake showed through a break in the trees, well below us; the wind carried a distant mutter of surf. The office was empty. Three keys lay upon three registration cards, each labeled with a customer's name. A sign said "Please fill out the registration card and leave it on the desk. We can settle the fee later."

As we were filling out our card the motel's owner walked in, accompanied by a young, extremely energetic collie. Within moments we were exchanging news of children and grandchildren, like old friends meeting after an emptiness of years. She and her husband would be flying out of Duluth the next weekend to attend the wedding of their son—their last unmarried child—someplace west of the Lakes. Minneapolis, maybe, or California. "I always hate to leave this place in someone else's hands," she said, "but you have to learn to trust people."

I indicated the cards and keys on the counter. "Looks like you've learned pretty well," I said.

She smiled. "When we bought this place," she said, "I made up my mind it had to be fun or I wouldn't do it. It's no fun being suspicious of people. And you know what? It works. Nobody's ever left without paying."

Back in Grand Marais we walked out Artist's Point, a place of polished rock and wave and woodland at the outer edge of the harbor. Large seas washed in and out over the tracks of ancient glaciers. Small winds and small birds sang in the conifers. Far out over the Superior Ocean squalls danced, their black clouds smiting the silver sea with electric forks and giving rise to rainbows. The teeth of the Sawtooth Range, as sharp and evenly spaced as their name suggests, faded southwest into golden mists.

We strolled to supper along the waterfront, to a place called the Blue Bay. Fifteen years before, Rod Badger and I had shared our last Great Lakes breakfast together here—a near-perfect 9.8—before driving downcoast to the airport at Duluth. The place had grown some since, taking over the second floor as well as the first of its old downtown building, but it was otherwise resolutely unchanged. It appeared, in fact, to be unchanged since approximately 1952. The food was simple and spectacularly good; the service, provided by a summer staff of college students from several foreign countries— Germany, France, China—was quick and friendly. We stepped out afterward into a bright wet world: a squall had come and gone, and the setting sun was gleaming within every beaded drop on the wet cars and wet buildings and wet lampposts.

Squalls continued to slide past through the night: rain spat at the roof, and the kettledrum-rumble of thunder mixed with a muffled undertone of surf wafting through the darkness from the shore a quarter mile distant. I slept fitfully, my mind on trip's end. Toward morning the squalls subsided, or perhaps drifted off, following the darkness westward. Pale light crept through the open window, carrying with it the whispers of waking birds.

THE PASSENGER FERRY TO THE WEST END
of Isle Royale leaves Grand Portage National Monument, from the
reconstructed dock used by the North West Company's fur traders in
the closing years of the eighteenth century, at 9:30 A.M. Passengers are
requested to be present an hour earlier, at 8:30: working backward,
with an allowance for breakfast and travel time, had us rolling out of
bed shortly after 6:00. As we drove into Grand Marais the sun popped
out of the sea, draping the dark conifers with baubles of bright gold.

The day before, we had sought our hostess's advice on breakfast
spots and had been directed to a small place called South of the
Border. "It's where the locals go," she said. "Tourists tend to avoid it
because of the name. They think it must be a taco shop." She smiled.
"Wrong border. People forget how close we are to Canada here." We
found the place easily, as much by the pickups with Minnesota plates
parked outside as by the small sign. Inside were the smells and sizzle
of good bacon and better coffee, and a companionable surprise: the
O.L.D.F.A.R.T.S., transplanted from beneath the Golden Arches and
looking remarkably healthy for the move.

At Grand Portage, the ferry—a sturdy little well-worn craft called
the *Wenonah*—was just nosing across the harbor toward the dock,
rolling a little with the swells. In the secured parking lot for the ferry
a young man carried a clipboard and a small box, checking off license
plates and collecting fees. A name tag identified him as official, but
not very effectively: cars were barely braking as they swerved by into
the lot. "This seems to be my day for being ignored," he said, as the
third car in a row whipped past without stopping.

"Wave your clipboard at them," I suggested.

"I tried that once. Had some guy ask me if I was doing a survey."
Another car came barreling by: the attendant looked at me and
shrugged wordlessly. Perhaps it was the incongruity of a parking
attendant where there was no town, but it seemed to me that people

using what was advertised as a secure lot might have expected someone there to secure it.

By the time we had walked to the dock the boat was pulled up and waiting, rocking gently, her white steel side groaning a little against the bumpers. In service since the 1960s, the *Wenonah* is sixty-three feet long and can hold up to a hundred people. For this trip she carried a crew of four. The captain, a slender, smiling man in his sixties, had been running the boat for about five years: this was his retirement project. His wife was the purser. Two deckhands, a master in his fifties and an apprentice in his early twenties, filled out the crew roster. The passengers were a mixed bag: backpackers headed to the island for a night or a week, families on day trips, an older man laden with photography gear, college students, couples. The captain's daughter and ten-year-old granddaughter were present, taking what appeared to be the granddaughter's first trip to the island. The captain came on the loudspeaker and warned us to cover our ears. Two sharp blasts on the diesel whistle and we cast off, backing away from the dock and cutting a smooth circle in the undulating surface of the harbor. Water and morning were both crystalline. A pretty teenager found a spot in the sun against the front of the wheelhouse and stripped to halter top and shorts. The apprentice deckhand suddenly found much to do in the prow. Under an air force of gulls the *Wenonah* cleared the bay mouth with its single large island and headed east, out onto the Superior Ocean.

On the open Lake the swells were running three to four feet, tall enough that their tops would occasionally curl over and break even in deep water. The boat cut across them at a slight angle, rocking and pitching. Some of the passengers turned pale. The teenager in front of the wheelhouse put on sunglasses and closed her eyes. Westward the land receded; eastward, Isle Royale was a vague line on the far side of a watery vastness. To the north lay the Susie Islands and, beyond them, the faint, distant forms of mountains on the Canadian mainland. To the south lay only the sea.

Isle Royale is nominally part of Michigan, but it is actually closer to Hudson Bay than it is to Detroit. Like many of the great Alaskan national parks—Glacier Bay, Lake Clark, Katmai—it is accessible only by boat or floatplane, and with its climate controlled by the cold Lake it looks and feels like Alaska. Its resident wolf pack and moose herd contribute to the northern ambience. The wolves and moose are the subjects of the world's longest continuous ecological study, a work that has been under way since 1958. In the summer they share space with park visitors. In the winter the park shuts down, and visitors are not permitted. From November to April the only large inhabitants of the island are the wolves and moose and the scientists who study them.

The island is forty miles long and eight wide; it is composed of a series of long parallel ridges which trend northeast-southwest and reach as much as seven hundred feet above Lake level. Bits of the ridges emerge offshore, surrounding the main island with a fleet of islets lined up in rows, as if standing at parade rest. At the northeast end, where the ice lingered longest, there is much bare, polished rock and numerous small lakes; the land is largely covered with spruce and fir, the trees of the taiga. Most of the southwest end is buried beneath outwash deposits, and although the land is actually higher here the topography is softer. Thicker soils and slightly warmer temperatures have favored higher concentrations of hardwoods; there is much maple and beech. Inland lakes are nearly nonexistent. The main exception, Feldtmann Lake, is exceptional also in its means of formation: it is an isolated embayment of Lake Nipissing, Superior's immediate predecessor. The barrier beach that cut it off from Nipissing is still visible at the lake's western end, high and dry and half a mile from the current Superior shoreline.

In 1983, Rod Badger and I came to the island with a canoe and spent three days exploring the waterways and trails of the east end. We camped on polished Precambrian volcanics at Moskey Basin, hiked the trail to Lake Richie, climbed Mt. Franklin. It rained, then cleared. Loons and wolves called in the night. Dragonflies the size of sparrows looped and dove through the dawn and twilight, sweeping mosquitoes from the air like an insect air force sent to provide cover for our invasion.

Melody and I had no canoe in 1998, and looming commitments at home: we could spare only a day for Isle Royale. That really meant just two and a half hours: the *Wenonah* would arrive at Windigo, on the island, at about 12:30 P.M., and depart for the mainland again at 3:00. There would be time for little more than a taste, not the full meal of fifteen years before. But a taste of Isle Royale trumps a lifetime in most of the rest of the country. I deeply regret that we did not have more time to spend on the island. I do not regret—and will never regret—that we came to the island at all.

Shortly past noon the *Wenonah* nosed her way through North Gap, the narrow passage between Isle Royale and Thompson Island, and turned left, up the long, slender entrance to Washington Harbor.

The swells subsided; a white seagull circled in sunshine before an intensely green wall of trees. Near the harbor's head the way split around high-humped Beaver Island. The captain took the left-hand path. A reef held cormorants. Beyond the height of Sugar Mountain— named for its maples—tall white cumuli gathered, promising moisture by midafternoon.

Windigo was a cluster of small worn buildings straggling up a hill beside the harbor. Hefting knapsack and cameras, we headed out the Feldtmann Lake trail. The trail passed through deep, delicious woods along the edge of the harbor. The leaves of queen's cup and dogwood bunchberry made a green false floor a few inches above the soil; their clustered berries—white for the queen's cup, red for the dogwood—stood up on short stems. Ferns and fungi abounded, the fungi themselves supporting furry mousecarpets of moss. In sunny spaces, asters bloomed like purple pinwheels.

A third of a mile or so from Windigo, a short side trail led down a bank to tiny Second Beach. We found a log and spread out our lunch. In the distance the *Wenonah* stood at her dock, a graceful white shape against dark green conifers. Superior lapped at our feet. Nearby, in a tiny cove, a mother merganser led six nearly grown offspring through what appeared to be diving lessons. In the wood near the water, a pile

Dogwood bunchberry, Isle Royale National Park, Michigan

of boards we first took for a tumbledown cabin turned out, instead, to be an ancient dory, pulled up and abandoned. What remained of the boat's sides had turned silver: there was no bottom. Saplings and small trees emerged from the interior like the ghosts of standing fishermen. Later, back at Windigo, I asked about the boat and received mostly blank stares. One staff person had seen it. No one knew where it had come from.

We walked another mile or so toward Feldtmann Lake. The trail crept back a bit from the harbor but did not otherwise change character. The forest was green and dim and felt much like that of Puget Sound. This was not entirely illusory: Puget Sound and Isle Royale came out from under the ice at roughly the same time, and they share a common water-dominated geography and a common flora. The dogwood bunchberry and queen's cup were familiar friends; so were most of the ferns and fungi, and the green mounds of thimbleberry, and if the trees were not of the same species as those in the West, they were at least of the same genera. Elsewhere in the Isle Royale archipelago the similarities to Puget Sound are even stronger. On certain small islands near Rock Harbor, too-casual visitors may find themselves entangled in devil's club, a spiny horror of the ginseng family which—thankfully—grows *only* here and in the vicinity of Superior's sister body of water in western Washington.

Shortly after we turned around it began to rain, lightly but insistently. Melody draped herself in a poncho; I unfurled the small umbrella I carry in my camera bag and hoisted it protectively over the cameras. Thus accoutred, we ran smack into most of the rest of our *Wenonah* shipmates, trailing obediently through the rain behind a park naturalist. A woman with soaked hair straggling down her shoulders pointed at my umbrella. "No fair!" she said sternly.

"Smart!" I replied.

She broke into a damp grin.

With half an hour to go before boarding, we strolled uphill to a small fenced area known as the Moose Exclosure. The rain had stopped, but the woods were wet and dark. The Moose Exclosure was darker still. The fence keeps moose out but lets all other creatures through; the area within it provides a taste of what Isle Royale would be like without its big vegetarian bulldozers. The difference is remarkable. The skinny stems of the trees crowd close together, like spectators at a popular horse race; undergrowth mounds over their feet like mutant dust bunnies. Little light penetrates. The place is oppressive, as if something had gone horribly wrong. In fact, something has. Large browsers are a necessary part of a forest. North American woodland plants are adapted to the presence of *Cervidae,* the family of mammals that includes deer, elk, and moose; their growth rates are adapted to cervid browsing habits. Without browsing they grow too fast, and too prolifically. Plants elbow their neighbors, shade each other out, stunt each other's growth. A browserless wood is choked, ugly, and impassable. It may be entirely "natural." It is not the type of nature anyone would go out of their way to see.

It is, of course, possible to have *too much* browsing. That is where predators come in. In the same way that woodland plants are adapted to large browsers, large browsers are adapted to predators, and when the predators disappear the browsers, like the browserless woods within the Moose Exclosure, grow too prolifically for their own good. That is the missing ingredient in a place like Ohio's Cuyahoga Valley National Recreation Area and Marquette's Presque Isle Park. We have chased the large predators away from these urban wildernesses because they are too dangerous to us, and then—because we are an empathetic species, and dislike the deaths of others—we have refused to take their place and kill the big browsers ourselves. The result is a browser population that is too big and an ecosystem that suffers. When our love of deer leads us to prohibit hunting, it is an Oscar Wilde love, a love straight out of *The Ballad of Reading Gaol:*

> Yet each man kills the thing he loves,
> By each let this be heard,
> Some do it with a bitter look,
> Some with a flattering word,
> The coward does it with a kiss,
> The brave man with a sword!

In the Cuyahoga and at Presque Isle, we have not been brave enough. As a society, we must understand that our kiss is killing the deer anyway, and the woods are dying with them. Better the sword.

Here on Isle Royale the sword is wielded by wolves, though not in a consistently effective manner. As big as it seems, the island is too small to support a steady wolf population; there is not enough food for a sustained series of good reproductive years, and no source of outside recruitment for a sustained series of bad ones. Thus the population spikes up and down. The moose population also spikes up and down, in a cycle only loosely tied to that of the wolves. Harsh winters and the moose's own profligacy probably play larger roles than wolf population dynamics do. When food is plentiful, the moose multiply; when it is scarce, they starve, and the population size plunges. Wolf predation acts as a governor on this cycle, keeping the spikes smaller, but it does not drive it. The wolves have only been a factor since they arrived on the island in 1947. The moose cycle has been going on since the moose got here fifty years before.

The relatively recent arrival dates of both wolf and moose on Isle Royale—they got here well after Europeans did—emphasize the role that change plays, even in a naturally driven ecosystem such as this. Both species arrived on their own, the moose by swimming, the wolves by crossing the fifteen-mile-wide ice bridge from the Canadian mainland that forms during particularly harsh winters. The flora of the island has been adapting ever since. Neither the inside nor the outside of the Moose Exclosure represents the pre-contact condition of Isle Royale. When Europeans first got here, a different large cervid, the woodland caribou, inhabited the island. The moose drove them off. Moose and caribou like different foods (a biologist would say they have "different selective browsing preferences"), so things the caribou ignored the moose have browsed to near-extinction. At the same time, things caribou like but moose don't have been thriving. This part of the island, around Windigo, was a small copper-mining settlement in the late nineteenth century. The forest the copper miners knew had a thick understory of yew and ponds full of water lilies. Today there are few water lilies and almost no yew, but there are lots of thimbleberries, a food which caribou like much better than do moose. The trail we took along the harbor was shaded by a canopy containing much old balsam fir. There is little new growth of balsam fir on the west end of Isle Royale today: whenever a seedling shows up, the moose eat it. The old trees are dying, and there are no young trees to replace them. The firs appear to be holding their own on the east end of the island, but on the west end they are expected to be functionally extinct by approximately 2010.

But perhaps I am attending to the wrong things. The most important feature of Isle Royale National Park is its wildness, and that is not changing. Only the unimportant details alter. Back in Houghton a week ago, we stopped at the park's headquarters and spoke with the ranger on duty at the shoreward visitor center. When I asked him if the park had changed much, he looked thoughtful. "A man and his young son came through here not long ago on their way out to the island," he said. "The man was asking me about a particular rock in Lake Richie—he wanted to know if it was still there. He had visited Isle Royale at the age his son was now, and this one rock in the lake had evidently meant a lot to him. Of course, there are lots of rocks in Lake Richie. I had no idea what to tell him." The ranger paused and smiled. "On their way back they stopped again, and the father said he had found the rock. He said it was completely unchanged. Now, we know water levels at Lake Richie have gone up and down, and the vegetation around it has changed. The water was full of PCBs from aerial deposition for a while, and now the PCBs appear to be covering over. There've been *lots* of changes. But to that father, nothing had changed. His boy got to see the same rock he did. He had tears in his eyes, right here in the public area. I'd say that's pretty good testimony."

When the *Wenonah* nosed out of Washington Harbor again onto the open Lake, the seas had increased even more: we found ourselves rollercoastering over swells four to six feet high, with an occasional maverick as high as eight feet. The boat strained up the swells and

The *Wenonah* docked in Washington Harbor, Isle Royale National Park

plunged down their far slopes, rolling from side to side. Strain; plunge: strain; plunge. The bigger swells slapped the prow, sending massive amounts of spray flying onto any passengers who happened to be on deck. Rain began again, although it was difficult to separate its effects from those of the spray. Melody, who suffers somewhat from motion sickness, buried herself in her poncho and went to stand in the spray: by facing forward and keeping her knees loose she managed to stave off the malady and even have fun for the first two hours, although she said later that the last hour got a little old. Others were not so fortunate: many spent much of the homeward journey hanging over the rail. The apprentice deckhand moved among them, his face a mask of resignation, passing out barf bags. Between these missions of mercy he sat in the aft cabin playing with a handheld electronic game. Early in the return voyage I had noticed him hitting mildly on the pretty teenager from this morning, but that had evidently not got very far. I asked him if he had been crewing long.

"Nah," he replied. "Just this summer."

"Do you expect to do this for a living?"

"I doubt it."

"Planning a Park Service career?"

"Nope."

"Just like boats?"

"Nope."

I was mystified. "So why did you take this job?"

He grinned suddenly. "My girlfriend conned me into it," he said.

At the ship's tiny rolling snackbar I ordered a cup of coffee and, using a trick I learned long ago from a jet boat operator in Idaho's Hells Canyon, set it firmly on the counter between sips, my hands around it to keep it from sliding away. A ship's motions are actually less likely to spill liquids than human motions are as we try to compensate for them. Midway through the coffee someone joined me at the counter: I turned, and saw that it was the captain.

"Who's driving the boat?" I asked.

"I don't know," he smiled. "Is someone supposed to be driving?"

He had left the wheel in the hands of the master deckhand, a man with as much or more experience as the captain himself, and had come down for coffee. I noticed he was spilling it on his hands. "Put it on the counter," I suggested.

He looked at me blankly.

"People motion," I said. "That's what spills coffee. A man named Ernie Duckworth taught me that on the Snake River."

The captain put his cup on the counter. "I'll be darned," he said after a moment. "It works."

Outside, the big Lake slapped at the boat, tossing it from hand to hand like a berserk juggler. Water stretched away to the south like a great rumpled bed, rolling on its springs, rolling us with it, unwilling lovers but unable to disengage. Rain hid the horizon, which would have been nothing but water and sky anyway. A young man passed by on his third trip to the head. "Difficult to do under these conditions," he mumbled.

The captain had the wheel again. The master deckhand joined me briefly at the port rail. "Having fun?" he asked.

"Actually, I am," I said. "I'm enjoying myself."

"Of course you are."

He passed behind me and then leaned back from the left, leering like a B-movie Nazi. "You haff no choisse," he growled, and was gone.

As we neared Grand Portage the rain spun off and the sun appeared low in the West, giving birth to perhaps the brightest rainbow I have ever seen, its head in the clouds, its base on the water among the black silhouettes of the Susie Islands.

> God gave Noah the rainbow sign;
> No more water; fire next time.

Actually, fire had its turn with this land eons ago, when the stone of the shore was shaped; it was water that was our benediction. The Superior Ocean, as limpid as birdsong and as big as God, endlessly blessing us, full of cold and promise as a winter dawn. We had found the Misty Isles, and the gods who dwelt there were wiser, even, than children; truer than any tale we could tell of them. Superior, as much as the rainbow, was the sign.

Back on the land, legs still jittery, we sought the car. Soon we were driving south, in sunshine. The blue Lake stretched east behind birch trunks that glowed yellow in the late light. South of Hovland, Highway 61 swept down beside a small open bay where waves rushed in past two large flat-topped rocks and rolled up a red beach covered with angular waterworn pebbles. I braked to a halt in a hail of sudden memories. Here Rod and I had stopped, in the cool balm of an early morning in August, and put the canoe in the water for the last time. Superior had been in one of her rare quiet moods that day, with swells running between six and nine inches. We paddled out to the rocks, considered landing on them, thought better of it. Back at the beach we pulled the boat up parallel, preparing to disembark, as we had done without incident at least two dozen times in the last three weeks. This time one of the swells caught us. The canoe came up on its gunwale and very nearly pitched us into the Lake. "What a way to end the trip!" Rod snorted as we righted. "We look like a pair of klutzes." We lashed the boat back on the car and left. Four hours later I was putting him on a plane in Duluth.

More in the past than the present, I prowled the red shore. Melody sat and watched. The ghost of a green canoe came riding in on a wave, turned broadside to the beach, and lost itself in a shower of sunlit spray. There was nothing more to see. We got in the Escort and fled south across the Kadunce River—called "Diarrhea Creek" by early settlers for what drinking its water will do to you—toward a small yellow motel nestled in pines, by Grand Marais, in the twilight.

VIII
FAREWELL

Our last dawn on the Lakes came quietly, on the backs of twittering birds. The mutter of surf had stilled. Grand Marais was just coming awake as we passed through, its harbor silver beneath a breath of mist. Gulls sat quiet as deacons. On the small hill southwest of town, maples bathed themselves in the sun. The summer trembled on the edge of autumn.

At the Cascade River we stopped and went down to the beach. Clouds hovered over the land, but seaward the air was full of light. The beach was bookended by outcrops of smooth basalt, its irregular surface holding splash pools rimmed with gold by the slanting sun. The Lake licked quietly at its pebbled shore, rising and falling perhaps three inches. It seemed hardly possible that this could be the same body of water whose six-foot swells had tossed us around like mice in a washing machine only a few hours before.

What did I learn in my lakeshore month? What song of the Lakes shall I sing as the days grow dark next winter? What have I learned on the coast of the Superior Ocean through the turning of days and years? (*UpLake, offshore rocks stand in silhouette, black against bright gold mists. Small swells nuzzle their flanks like hungry kittens. There is a smell of spruce and water.*)

Let us reach back to the first trip and the book that came of it, and compare.

Fifteen years ago I was frantic about lakewide pollution. That fear has been substantially calmed, though not thoroughly laid to rest. Lake Erie is cleaner than it has been for a couple of generations. Places whose names were synecdoches for environmental horror— Waukegan Harbor, Hamilton Harbour, the Cuyahoga River—have been scrubbed substantially and, if they do not sparkle, at least no longer terrify. Love Canal has been capped as meticulously as a starlet's molars. To claim that the Lakes are clean would be overstating the case: there is still the Fox River, and the harbor at Ashtabula, and

A storm over the Sawtooth Range from Artist's Point, Grand Marais, Minnesota (photo by Melody Ashworth)

the demons that haunt the lower reaches of the Grand Calumet. The dream of zero discharge, put forth so hopefully in the Great Lakes Water Quality Agreement twenty years ago this summer, remains something we see only with our eyes closed. Still, it would take a small-minded person, indeed, to deny that substantial progress has taken place. The focus has shifted from cleaning up the Lakes to preventing backsliding while cleaning up the remaining corners. The task is still substantial but the trend is positive, and all Great Lakes residents I spoke with remained firmly committed to making it happen.

A second concern *The Late, Great Lakes* focused upon was diversions and consumptive uses, and here the outlook is somewhat less sanguine. Massive pipelines to the parched West remain the phantoms they have always been; thus far economics, not environmental protest, has killed every one. The High Plains, where the water is needed, stand as much as four thousand feet above the level of Lake

Superior, and pumping water that far uphill is prohibitively expensive. But the near-morbidity of the big-pipe scenario has not stopped the proliferation of small pipes. Besides the long-standing Chicago Diversion, there are diversions existing or proposed at two places in Wisconsin—Crandon and Pleasant Prairie—and at least one in Ohio, at Akron. There are numerous other places that are eyeing diversions, and some that have quietly done it—by pumping groundwater, for instance, from wells whose cones of depression reach inside the Great Lakes Basin, short-circuiting aquifers which would otherwise feed springs that run into the Lakes. There are proposals to run water downhill from Lake Huron to feed the proliferating northern suburbs of Toronto; once used, the water would become sewage, which would be treated and discharged into Lake Ontario, thus bypassing the St. Clair/Detroit River system, Lake St. Clair, Lake Erie, and Niagara Falls—all of which would suffer from reduced water supplies. Finally, there is the looming specter of out-of-basin sale of bottled water. It nearly happened this summer, and we certainly have not seen the end of it. Communities all over the world are paying significant sums to clean up their water supplies; residents of the Great Lakes Basin are sitting on a supply that is still naturally clean. Recall Bob Beltran's statement: "You can sit out in the middle of any of the Lakes, even Lake Erie, and drink to your heart's content." Most of the world cannot do that. It is going to become increasingly difficult to justify denying them the chance to do it secondhand. The Lakes could still be drained: pumping water to the High Plains could yet prove feasible as clean water becomes more scarce. Continued vigilance is necessary.

Problems with exotic species have shifted dramatically and have probably worsened, although that is a subjective judgment: who is to say that the current plague of zebra mussels is worse than the plague of alewives that created the stink of small black bodies I saw on beaches in 1983, or the plague of lampreys that preceded those? Alewives are still around, but current efforts aim to manage them as forage fish for salmon rather than to eliminate them as a nuisance. (The fact that the salmon themselves are planted exotics is interesting but probably not significant, at least so long as they continue to fail to spawn.) Smelt, likewise, have settled in, to the point where most Great Lakes residents are probably unaware that they were once considered a nuisance on a par with the zebra mussel. Lamprey have certainly *not* settled in, and continue to be kept under control only by strenuous effort. It is worth noting, however, that the Great Lakes have four native lamprey species in addition to the sea lamprey. Two of the four natives are parasitic in the same manner as their exotic cousins. Neither has been a problem, at least in historic times. Is it possible that our insistence on controlling the sea lamprey through chemical means has prevented biological controls from developing? The question is at least worth asking. Though it has proved difficult for us to keep in mind, ecosystems are driven, not by species, but by niches. The introduction of an exotic creates new niches and empties old ones, and while these changes are taking place things can go enormously out of whack. But niches always fill. It is merely a matter of remaining patient.

While we are waiting, of course, things can get pretty grim. The immense piles of zebra mussel shells we saw on the beaches of Erie were not a pretty sight. Here on Superior, zebras remain rare—Park Service employees told me they have never seen one on Isle Royale, and Bill Robinson had to have some mailed to him in Marquette— but the Superior Ocean has the river ruffe and the round goby. All the Lakes have problems with the spiny water flea. Everywhere, wetlands have sprouted the tall lavender tongues of purple loosestrife. The mechanics of niche competition will eventually flatten these out, but in the meantime we may lose a number of things we love. The cost of waiting may be more than we can afford. We cannot always wait for biology. But biology cannot wait for us, either. Exotic species are always going to be a problem. Perhaps we should stop spending so much effort trying to eradicate them and give a little more attention to how to live with them.

Thinking about exotic species brings up the question of biodiversity. (*There is a grebe out there, or possibly a loon: it is difficult to tell which at this distance. What I see is a small black silhouette, long-necked and low-backed, disappearing as it dives and appearing again some random time and distance away, the water about it bright with still-early sun.*) Biodiversity was not a word that was spoken much in 1983, but it was really the heart of the issue, then as now. In common with all places, the Great Lakes are not a single system, but a mosaic of small, interconnected ones. Each time a connection is severed—each time a small system dies—the mosaic is weakened. In time it will fail. We have made great strides in understanding this. Small systems are now being identified (alvars), protected (Old Woman Creek; the Mink River Estuary; the Chicago Wilderness), and carefully studied. The Kent Fullers and Jane Elders and Eddie Herdendorfs and Bill Robinsons of this world are continuing to look out for the rest of us. They are still outgunned by those who do not understand, or who do not wish to; but the balance is shifting. People are beginning to realize that a healthy environment is not the adversary of a healthy economy, but a necessary precondition for it.

Of course, it is one thing to understand, and quite another to act on that understanding. No person I met on this trip understands the dangers inherent in shoreline development better than Eddie Herdendorf, who maintains a shoreline house anyway. Kent Fuller, who can go on for hours about the problems caused by second homes in northern Wisconsin, has a second home in northern Wisconsin. To say this is not to condemn the actions of either man; it is simply to recognize the seductive force of the Fallacy of Composition. Building homes in the woods or along the shore is logical and desirable, but each one contributes to the illogical and undesirable situation we find ourselves in today. Everywhere we went on this trip—almost literally everywhere—there was sprawl. The more attractive the place, the worse the problem. The old enclaves, like Amy Blossom's Castle Park, are still islands, but the sea surrounding them is no longer the faceless wild; it is the faceless suburbs. It is not necessary to take the elitest stand that the door should have been shut after the Blossoms and the Fullers and the Herdendorfs got there to see that surely the door must shut *sometime*, if only when all the available land is gone. If we are wise, as a people, we will shut it before then. Otherwise the places we are escaping to will become just like the places we have left, and there will no longer be any point.

But to shut a door and keep it shut—especially against as insidious and persistent a visitor as the Fallacy of Composition—requires government action: and unfortunately this is the precise spot where we are most blatantly failing. I am no longer a fan of command-and-control legislation: it has proved too rigid, and it often prevents nearly as much environmental progress as it promotes. Standards are boundaries, which makes them a poor fit to the boundless world of nature. To concede this, however, is not the same thing as advocating the elimination of controls altogether. We need to focus government differently, not starve it. Too often we see legislators elected on a platform that pledges the fiscal equivalent of anorexia: no matter how skinny and weak government becomes, it continues to look at itself through legislative eyes and pronounce itself too fat. We certainly do not need a government that throws its weight around like a schoolyard bully, but we do need one that is healthy enough to hammer the bullies who come after us.

Here on the Great Lakes, the most glaring example of damage

by government downsizing is the withdrawal of funding for the RAPs. RAPs are not command-and-control tools: they are flexible processes, focused on solutions rather than problems, inclusive—the Public Advisory Councils cover a broad range of viewpoints—and amazingly successful. They are a principal reason why pollution, the Lakes' worst problem in 1983, is mostly a minor nuisance today. But everywhere around the Lakes, on both the U.S. and Canadian sides, budget cutters are slashing support for them. They are victims of their own success. Pollution is now such a low-profile issue that politicians can get away with ignoring it. Oversight wastes away, and the vultures fledge their young for the next foul round.

Sprawl is also a failure of oversight: but in this case it has not been withdrawn, it has simply never developed. We are a culture born of open spaces, and though Frederick Jackson Turner warned us of their disappearance over a century ago we have yet to make the necessary adjustments. There is an old axiom that goes: *Your freedom to swing your fist ends when it encounters my face.* We are now fist to face all along the lakeshore, and it is time to stop swinging. We pay for suburban sprawl three times: first, through the loss of open space and biodiversity; second, through subsidy of the infrastructure— waste treatment, water supply, roads, power lines, phone lines— that supports new development; third, through the reduction of infiltration capacity, interruption of shore drift, and other physical changes that come when a land is pinned down too tightly by houses and can no longer dance freely to the ancient rhythms of wind and water. Here, not through taxes, is where the wealth is truly being drained from us. If people are going to continue to demand the right to build wherever they please, let us at least make them cover the full cost. "We have become a people that has turned its back to the sea," I wrote in *The Late, Great Lakes*, "and the consequences—for sea and people alike—are grave." Happily, this statement is no longer true. In fifteen years, we have learned to see the Lakes as the treasure they are. Apathy appears to be dead. Now we must deal with the fallout from too much attraction.

In midmorning we stopped at Tettegouche and hiked the half-mile trail to Shovel Point. The trail climbed through maples twitterfull of warblers to a tall headland where 180-foot cliffs dropped sheer to the sea and the constant sweep of winds had created a dwarf forest like those at the northern treeline. At the open tip of the point, where the trail descended to near-Lake level above rock scoured clean by the surf, there was a murmur of voices. Members of an encounter group of some sort had left the trail and were squatting in a solemn circle in the shrubbery a hundred feet or so away, their packs flat beside them. This, too, is a legitimate use of the shore, but it is one that Radisson and Groseilliers certainly never encountered. We cut our own stay short and made our way back to the warblers. Popularity is an insidious disease.

Just before noon we stopped one final time. There is a small park called Kitchi Gammi on the outskirts of Duluth, and it is here, on every trip to Superior so far, that I have come to say good-bye. A wide flat outcropping of ancient rock stretches seaward a few feet above water level, scoured smooth by glaciers and by centuries of restless surf: on days of big waves the power of the ocean is here, the heavy water roaring like Jehovah as it smashes to crystal shards against the immobile stone. This time, uncharacteristically, the water was still; stiller, even, than it had been at Cascade River. The great Lake seemed to be holding its breath. Tiny swells, no more than half an inch high, made small talk with the rocks. I crossed via nearly submerged stepping-stones to a rock perhaps four feet square a short distance offshore and sat for a while, scanning the flat horizon for signs and wonders, and while I was there the Lake exhaled slightly. The half-inch swells increased to four-inchers; the stepping-stones I had used to reach my perch went under. I took off my shoes and socks and waded ashore, the water waving good-bye against my calves. Back on the beach I sat on smooth stone and dried my feet in the sun, looking one last time toward the distant edge of the world. Then we left.

We climbed through Duluth to the Basin's rim. The continent stretched westward, flat and green in the long afternoon. The sun fell toward Oregon. We followed.

Farewell, Superior, until we meet again.

Bibliography

1000 Islands International Travel Guide. Alexandria Bay, N.Y.: 1000 Islands International Tourism Council, n.d.

1982 Report on Great Lakes Water Quality. Windsor, Ontario: Great Lakes Water Quality Board of the International Joint Commission, November 1982.

1983 Report on Great Lakes Water Quality. Windsor, Ontario: Great Lakes Water Quality Board of the International Joint Commission, November 1983.

1987 Report on Great Lakes Water Quality. Windsor, Ontario: Great Lakes Water Quality Board of the International Joint Commission, November 1987.

"The 1998 Weyerhaeuser Au Sable River Canoe Marathon July 24–25 Spectator Guide." Weyerhaeuser Au Sable River Canoe Marathon Web site (http://www.ausablecanoemarathon.org/specta98.html), accessed February 17, 1999.

Apostle Islands Official Map and Guide (brochure). Bayfield, Wis.: Apostle Islands National Lakeshore, National Park Service, 1998.

Around the Archipelago: Apostle Islands National Lakeshore (brochure). Bayfield, Wis.: Apostle Islands National Lakeshore, National Park Service, 1998.

Au Sable National Scenic River (brochure). Mio, Mich.: Mio Ranger District, Huron-Manistee National Forest, U.S.D.A. Forest Service, n.d.

"Au Sable River Canoe History." Oscoda.Net Web site (http://www.oscoda.net/arcm/canhist.html), accessed February 17, 1999.

Battin, J. G., and J. G. Nelson. *Man's Impact on Point Pelee National Park*. Toronto: National and Provincial Parks Association of Canada, 1978.

"Bay of Quinte Improvements in Management." *The Green Lane*. Environment Canada Web site (http://www.ec.gc.ca/pp/english/stories/quinteen.html), accessed March 24, 1999.

The Beautiful Bruce Peninsula, Ontario, Canada. Wiarton, Ontario: Bruce Peninsula Tourism, n.d.

Beauty and the Beast: Purple Loosestrife (brochure). Indiana Dunes National Lakeshore, National Park Service, n.d.

Besser, John M., et al. "Assessment of Sediment Quality in Dredged and Undredged Areas of the Trenton Channel of the Detroit River, Michigan USA, Using the Sediment Quality Triad." *Journal of Great Lakes Research* 22, no. 3 (September 1996): 683–96.

"Bill Summary and Status for the 105th Congress: H.R. 2400, Transportation Equity Act for the 21st Century." *Thomas— Legislative Information on the Internet*. Library of Congress Web site (http://thomas.loc.gov/cgi-bin/bdquery/z?d105:h.r.02400:), accessed April 10, 1999.

"Bill Summary and Status for the 105th Congress: S.1408, D'Amato, Establishing the

Lower East Side Tenement National Historic Site." *Thomas—Legislative Information on the Internet.* Library of Congress Web site (http://thomas.loc.gov/cgi-bin/bdquery/z?d105:SN01408:@@@L), accessed April 10, 1999.

Blue Mound State Park Visitor (brochure). Blue Mound, Wis.: Blue Mound State Park, Wisconsin Department of Natural Resources, 1997.

Bogue, Margaret Beattie, and Virginia A. Palmer. *Around the Shores of Lake Superior: A Guide to Historic Sites.* Madison: University of Wisconsin Sea Grant College Program, 1979.

Boyer, Barry. "The Faltering Search for Community in Great Lakes RAPs." *Journal of Great Lakes Research* 23, no. 2 (June 1997): 229–31.

———. *No Place to Hide? Great Lakes Pollution and Your Health.* Buffalo: Baldy Center for Law and Social Policy and the Great Lakes Program, State University of New York at Buffalo, 1991.

Bredin, Jim, ed. *Great Lakes Trends: A Dynamic Ecosystem.* Lansing: Office of the Great Lakes, Michigan Department of Environmental Quality, 1998.

Brockwell-Tilman, Elizabeth, and Earl Wolf. *Discovering Great Lakes Dunes.* Muskegon, Mich.: Gillette Natural History Association, 1998.

Bruce Peninsula National/Fathom Five National Marine Parks. Tobermory, Ontario: Bruce Peninsula National Park and Fathom Five National Marine Park, Canadian Heritage Parks Canada in conjunction with the Tobermory Press, 1998.

Bruce Peninsula National Park (brochure). Tobermory, Ontario: Bruce Peninsula National Park, Environment Canada Parks Service, 1992.

Cantor, George. *The Great Lakes Guidebook.* 3 vols. Ann Arbor: University of Michigan Press, 1978, 1979. and 1980.

Cape Vincent: Gateway to the Thousand Islands, the Golden Crescent, and the Beautiful St. Lawrence River. Cape Vincent, N.Y.: Cape Vincent Chamber of Commerce, 1998.

"Chaumont Barrens." *Nature Conservancy Preserve Profiles.* The Nature Conservancy Central and Western New York Chapter Web site (http://www.tnc.org/infield/State/NewYork/central/profiles/barrens.html), accessed March 22, 1999.

Chaumont Barrens: Jefferson County's Hidden Ecological Treasure (brochure). Rochester: Central and Western New York Chapter, The Nature Conservancy, n.d.

"Chicago Area Paddling/Fishing Guide: Waukegan Harbor (Lake County, Illinois)." (http://pages.ripco.net/jwn/waukegan.gif), accessed December 18, 1998.

Chutes Provincial Park (brochure). Massey, Ontario: Chutes Provincial Park, Ontario Ministry of Natural Resources, n.d.

"Coalition Seeks Oil and Gas Development Planning." *Great Lake Bulletin,* Spring 1997. Michigan Land Use Institute Web site (http://www.mlui.org/pubs/glb/glbsp97/GLBsp9719.html), accessed February 16, 1999.

Corkum, Lynda D., Jan J. H. Ciborowski, and Rodica Lazar. "The Distribution and Contaminant Burdens of Adults of the Burrowing Mayfly, *Hexagenia,* in Lake Erie." *Journal of Great Lakes Research* 23, no. 4 (December 1997): 383–90.

Coscarelli, Mark. *Nonindigenous Aquatic Nuisance Species State Management Plan: A Strategy to Confront Their Spread in Michigan.* Lansing: Office of the Great Lakes, Michigan Department of Environmental Quality, January 1996.

"Council of Great Lakes Industries Position Statement on the Remedial Action Plan Process." *Journal of Great Lakes Research* 23, no. 2 (June 1997): 225–26.

Craigleith Provincial Park (brochure). Collingwood, Ontario: Craigleith Provincial Park, Ontario Ministry of Natural Resources, n.d.

Crane Creek Wildlife Research Station (brochure). Oak Harbor, Ohio: Crane Creek Wildlife Research Station, Ohio Division of Wildlife, 1995.

Daniel, Glenda, and Jerry Sullivan. *The North Woods of Michigan, Wisconsin, Minnesota: A Sierra Club Naturalist's Guide.* San Francisco: Sierra Club Books, 1981.

"DEQ Denies Jordan Valley Drilling." News release, Michigan Land Use Institute, Benzonia, Michigan, May 9, 1997. Michigan Land Use Institute Web site (http://www.mlui.org/keyissues/projectindex/jordanrivervalley.html), accessed February 16, 1999.

"DEQ Hears Public on Jordan Drilling." News release, Michigan Land Use Institute, Benzonia, Michigan, January 10, 1997. Michigan Land Use Institute Web site (http://www.mlui.org/keyissues/projectindex/jordanrivervalley.html), accessed February 16, 1999.

De Vault, David S., et al. "Contaminant Trends in Lake Trout and Walleye from the Laurentian Great Lakes." *Journal of Great Lakes Research* 22, no. 4 (December 1996): 884–95.

Downs, Warren, and Christine Kohler. *Fish of Lake Superior.* Madison: University of Wisconsin Sea Grant College Program, [1976].

Downtown and Waterfront Tour Guide. Sault Ste. Marie, Ontario: Ontario Travel Center, 1996.

Dubs, Derek O. L., and Lynda D. Corkum. "Behavioral Interactions between Round Gobies (*Neogobius melanostomus*) and Mottled Sculpins (*Cottus bairdi*)." *Journal of Great Lakes Research* 22, no. 4 (December 1996): 838–44.

"Duluth Seaway Port Authority: Port Facts." Seaway Port Authority of Duluth. Duluth Seaway Port Authority Web site (http://www.duluthport.com/seawayportfacts.html), accessed April 19, 1999.

Eggold, Bradley T., James F. Amrhein, and Michael A. Coshun. "PCB Accumulation by Salmonine Smolts and Adults in Lake Michigan and Its Tributaries and Its Effect on Stocking Policies." *Journal of Great Lakes Research* 22, no. 2 (June 1996): 403–13.

"Environmentally Sensitive Organisms Missing in Lake Michigan Mud Samples." NOAA Office of Public Affairs Web site (http://www.glerl.noaa.gov/news/amphipods.html), accessed December 4, 1997.

Evers, David C. *A Guide to Michigan's Endangered Wildlife.* Ann Arbor: University of Michigan Press, 1992.

Exotics: Don't Let Them Ride with You! (brochure). Minnesota Sea Grant, University of Minnesota, n.d.

Farid, Claire, John Jackson, and Karen Clark. *The Fate of the Great Lakes: Sustaining or Draining the Sweetwater Seas?* Toronto and Buffalo: Canadian Environmental Law Association and Great Lakes United, 1997.

Frank, Andy. "Last Detroit River Marshland Is Imperiled." *Great Lakes United* 12, no. 2 (Spring 1998): 3, 28.

———. "United States Finally Moves against Exotic Species in Ships." *Great Lakes United* 11, no. 2 (Spring 1997): 9, 28.

Friends of Stone Laboratory Newsletter, Autumn 1996 edition. Columbus, Ohio: Friends of Stone Laboratory, 1996.

Froese, Kenneth L., et al. "PCBs in the Detroit River Water Column." *Journal of Great Lakes Research* 23, no. 4 (December 1997): 440–49.

Gamble, Teresa. "Bay of Quinte Remedial Action Plan, October 1997." *Great Lakes*

Information Management Resource (GLIMR). Environment Canada Web site (http://www.cciw.ca/glimr/raps/ontario/quinte/), accessed March 24, 1999.

———. "Collingwood Harbour Remedial Action Plan, October 1998." *Great Lakes Information Management Resource (GLIMR).* Environment Canada Web site (http://www.cciw.ca/glimr/raps/huron/collingwood/intro.html), accessed March 29, 1999.

———. "Spanish Harbour Remedial Action Plan, October 1998." *Great Lakes Information Management Resource (GLIMR).* Environment Canada Web site (http://www.cciw.ca/glimr/raps/huron/spanish/intro.html), accessed April 2, 1999.

Gilbert, Reg. "Holdout Michigan Approves Akron Water Diversion." *Great Lakes United* 12, no. 2 (Spring 1998): 1, 4, 28.

———. "Michigan Attacks Crandon on Groundwater Diversion; Hydrogeologist Blasts 'Misleading' Environmental Report." *Great Lakes United* 12, no. 1 (Winter/Spring 1998): 4, 14.

Gossiaux, Duane C., Peter F. Landrum, and Susan W. Fisher. "Effect of Temperature on the Accumulation Kinetics of PAHs and PCBs in the Zebra Mussel, *Dreissena polymorpha.*" *Journal of Great Lakes Research* 22, no. 2 (June 1996): 379–88.

Graham, J. Robertson. *Where Canada Begins: A Visitor's Guide to Point Pelee National Park.* Leamington, Ontario: Friends of Point Pelee, n.d.

The Great Lakes: An Environmental Atlas and Resource Book. 3rd ed. Chicago and Toronto: U.S. Environmental Protection Agency Great Lakes Program Office and the Government of Canada, 1995.

Great Lakes Commission 1995 Annual Report. Ann Arbor: Great Lakes Commission, 1996.

Great Lakes Piping Plover: An Endangered Species (brochure). Lansing: U.S. Fish and Wildlife Service, Region 3, April 1994.

Great Lakes Program Progress Report: United States Report to the International Joint Commission. Chicago: U.S. Environmental Protection Agency Great Lakes National Program Office, 1995.

Great Lakes Research Review, Vol. 2, No. 1: Great Lakes Fisheries Issues. Buffalo: Great Lakes Program, State University of New York at Buffalo, 1995.

Great Lakes Research Review, Vol. 3, No. 2: Great Lakes Exotic Species II. Buffalo: Great Lakes Program, State University of New York at Buffalo, 1998.

"Great Lakes Success Stories: Ed Houghton, Manager of Operations, Public Utilities, Collingwood." *Ontario Green Lane.* Environment Canada Web site (http://www.cciw.ca/green-lane/success-stories/gl/ed.html), accessed March 25, 1999.

"Great Lakes Water Quality Board Position Statement on the Future of Great Lakes Remedial Action Plans: September 1996." *Journal of Great Lakes Research* 23, no. 2 (June 1997): 212–20.

Green, John C. *Geology on Display: Geology and Scenery of Minnesota's North Shore State Parks.* Minnesota Department of Natural Resources, 1996.

The Greenstone: Isle Royale National Park Information (brochure). Houghton, Mich.: Isle Royale National Park, National Park Service, 1998.

Grey Bruce Visitors Guide, 1998. Owen Sound, Ontario: Grey Bruce Tourism Association, 1998.

Guest, Margery. "Of Time and the River: The Beginning of the Au Sable River

Canoe Marathon." *Michigan Natural Resources Magazine*, July–August, 1997. Weyerhaeuser Au Sable River Canoe Marathon Web site (http://www. ausablecanoemarathon.org/margery.html), accessed February 17, 1999.

Hartig, John H. "Great Lakes Remedial Action Plans: Fostering Adaptive Ecosystem-Based Management Processes." *American Review of Canadian Studies*, Autumn 1997, pp. 437–58.

Hartig, John H., and Neely L. Law. "Collingwood Harbour Area of Concern." *Progress in Great Lakes Remedial Action Plans, 1994*. Consortium for International Earth Science Information Network Web site (http://epaserver.ciesin.org/glreis/nonpo/nprog/aoc_rap/docs/progress/cover.html), accessed March 25, 1999.

Hartig, John H., Richard L. Thomas, and Edward Iwachewski. "Lessons from Practical Application of an Ecosystem Approach in Management of the Laurentian Great Lakes." *Lakes and Reservoirs: Research and Management*, issue no. 2 (1996): 137–45.

Hartig, John H., et al. "Quantifying Targets for Rehabilitating Degraded Areas of the Great Lakes." *Environmental Management* 21, no. 5 (1997): 713–23.

Hartwick Pines State Park, Grayling, Michigan (brochure). Lansing: Michigan Department of Natural Resources, Parks and Recreation Division, 1998.

Hartwick Pines State Park Old Growth Forest Foot Trail (brochure). Lansing: Michigan Department of Natural Resources, Parks and Recreation Division, 1998.

Havighurst, Walter, ed. *The Great Lakes Reader*. New York: Collier Books, 1978.

Hearing on H.R. 351, H.R. 1714, H.R. 2136, and H.R. 2283 before the Subcommittee on National Parks and Public Lands of the Committee on Resources, House of Representatives, One Hundred Fifth Congress, First Session, September 16, 1997, Washington, DC (Serial No. 105–51). Library of Congress Web site (http://commdocs.house.gov/committees/resources/hii45531.000/hii45531_0f.htm), accessed April 8, 1999.

Hill, John R. *The Indiana Dunes: Legacy of Sand*. Bloomington, Ind.: Department of Natural Resources Geological Survey Special Report 8, 1974.

Holcombe, Troy L., et al. "Lakefloor Geomorphology of Western Lake Erie." *Journal of Great Lakes Research* 23, no. 2 (June 1997): 190–201 and map insert.

Howell, E. Todd, et al. "Changes in Environmental Conditions during *Dreissena* Colonization of a Monitoring Station in Eastern Lake Erie." *Journal of Great Lakes Research* 22, no. 3 (September 1996): 744–56.

Hubbs, Carl L., and Karl F. Lagler. *Fishes of the Great Lakes Region*. Bloomfield Hills, Mich.: Cranbrook Institute of Science, 1947.

Important Health Information for People Eating Fish from Wisconsin Waters. Madison: Wisconsin Division of Health and Wisconsin Department of Natural Resources, 1998.

Indiana Dunes Cowles Bog Area (brochure). Porter, Ind.: Indiana Dunes National Lakeshore, National Park Service, n.d.

Indiana Dunes Official Map and Guide (brochure). Porter, Ind.: Indiana Dunes National Lakeshore, National Park Service, n.d.

Indiana Dunes Reservations of Use and Occupancy (brochure). Porter, Ind.: Indiana Dunes National Lakeshore, National Park Service, n.d.

The Indiana Dunes Story: How Nature and People Made a Park. 2nd ed. Michigan City, Ind.: The Shirley Heinze Environmental Fund, 1997.

Isle Royale National Park: Draft General Management Plan Environmental Impact Statement. Denver: National Park Service Denver Service Center, 1998.

BIBLIOGRAPHY

Isle Royale Official Map and Guide. (brochure). Houghton, Mich.: Isle Royale National Park, National Park Service, n.d.

Kellogg, Wendy A. "Lessons from RAPs: Citizen Participation and the Ecology of Community." *Journal of Great Lakes Research* 23, no. 2 (June 1997): 227–28.

———. "Metropolitan Growth and the Local Role in Surface Water Resource Protection in the Lake Erie Basin." *Journal of Great Lakes Research* 23, no. 3 (September 1997): 270–85.

Krantzberg, Gail. "International Association for Great Lakes Research Position Statement on Remedial Action Plans." *Journal of Great Lakes Research* 23, no. 2 (June 1997): 221–23.

———. "Research Priorities for Great Lakes Rehabilitation" (Editorial). *Journal of Great Lakes Research* 23, no. 3 (September 1997): 239–40.

Krieger, Kenneth A., et al. "Recovery of Burrowing Mayflies (Ephemeroptera: Ephemeridae: *Hexagenia*) in Western Lake Erie." *Journal of Great Lakes Research* 22, no. 2 (June 1996): 254–63.

Kuchenberg, Tom, and Jim Legault. *Reflections in a Tarnished Mirror: The Use and Abuse of the Great Lakes.* Sturgeon Bay, Wis.: Golden Glow Publishing, 1978.

"Lake Huron." Consortium for International Earth Science Information Network Web site (http://epaserver.ciesin.org/glreis/nonpo/nprog/aoc_rap/huron/), accessed April 2, 1999.

"Lake Huron Facts and Figures." The Great Lakes Information Network Web site (http://www.great-lakes.net/refdesk/almanac/lakes/huronfct.html), accessed February 20, 1999.

Lake Michigan Water Quality and You (brochure). Porter, Ind.: Indiana Dunes National Lakeshore, National Park Service, n.d.

Lakeport State Park, Lakeport, Michigan (brochure). Lansing: Michigan Department of Natural Resources, Parks and Recreation Division, n.d.

Lakeshore Observor (brochure). Munising, Mich.: Pictured Rocks National Lakeshore, National Park Service, 1998.

"Land Stewardship." *Detroit Free Press,* December 1, 1996: obtained from Michigan Land Use Institute Web site (http://www.mlui.org/keyissues/projectindex/jordanrivervalley.html), accessed February 16, 1999.

Lesko, Lynn T., Stephen B. Smith, and Marc A. Blouin. "The Effect of Contaminated Sediments on Fecundity of the Brown Bullhead in Three Lake Erie Tributaries." *Journal of Great Lakes Research* 22, no. 4 (December 1996): 830–37.

A Long Term Plan to Improve Modeling Capabilities for Toxic Chemicals in Lake Ontario (Donald W. Rennie Memorial Monograph Series Great Lakes Monograph No. 10). Buffalo: Great Lakes Program, State University of New York at Buffalo, 1998.

Lonky, Edward, et al. "Neonatal Behavioral Assessment Scale Performance in Humans Influenced by Maternal Consumption of Environmentally Contaminated Lake Ontario Fish." *Journal of Great Lakes Research* 22, no. 2 (June 1996): 198–212.

Madden, Anna Brooks, and Joseph C. Makarawicz. "Salmonine Consumption as a Source of Mirex in Human Breast Milk near Rochester, New York." *Journal of Great Lakes Research* 22, no. 4 (December 1996): 810–17.

"Manitoulin Island Ferry Service." The Original Tobermory Home Page Web site (http://www.kanservu.ca/arconst/ferry.htm), accessed March 31, 1999.

Marsden, J. Ellen, and John Janssen. "Evidence of Lake Trout Spawning on a Deep Reef in Lake Michigan Using an ROV-Based Egg Collector." *Journal of Great Lakes Research* 23, no. 4 (December 1997): 450–57.

McMillan, Ian, ed. *Algoma Outdoors.* Sault Ste. Marie, Ontario: Algoma Kinniwabi Travel Association, 1998.

Meyer, Michael, et al. "Endangered Shorelines." *Horizons* (Sigurd Olson Environmental Institute, Northland College, Ashland, Wis.), Summer 1998, pp. 7–8.

Michigan's Keweenaw Peninsula: A Guide to Selected Attractions, Lodging, Dining, and Shopping (brochure). Houghton, Mich.: Keweenaw Vacation, n.d.

Michigan State Parks Guide: Great Lakes, Great Times, Great Parks. Lansing: Michigan Department of Natural Resources, Parks and Recreation Division, 1998.

Michigan's Western U.P.: Clearly Superior. Ironwood, Mich.: Western U.P. Convention and Visitor Bureau, n.d.

Morrison, Todd W., William E. Lynch, Jr., and Konrad Dabrowski. "Predation on Zebra Mussels by Freshwater Drum and Yellow Perch in Western Lake Erie." *Journal of Great Lakes Research* 23, no. 2 (June 1997): 177–89.

Myers, Donna N., and Dennis P. Finnegan. "National Water-Quality Assessment Program—Lake Erie–Lake St. Clair Basin. *Water Resources of Ohio.* U.S.G.S. Web site (http://oh.water.usgs.gov/nawqa/index.html), accessed March 9, 1999.

Nalepa, Thomas F., et al. "Changes in the Freshwater Mussel Community of Lake St. Clair: From Unionidae to *Dreissena polymorpha* in Eight Years." *Journal of Great Lakes Research* 22, no. 2 (June 1996): 354–69.

Ninth Biennial Report on Great Lakes Water Quality. Windsor, Ontario: International Joint Commission, 1998.

Northern Great Lakes Visitor Center (brochure). Ashland, Wis.: Northern Great Lakes Visitor Center, n.d.

Nowhere Else on Earth Can You Find Wood Like This (brochure). Ashland, Wis.: Superior Water-Logged Lumber Company, Inc., n.d.

Old Woman Creek National Estuarine Research Reserve and State Nature Preserve (brochure). Huron, Ohio: Old Woman Creek Reserve, Ohio Division of Natural Areas and Preserves, n.d.

Oleszewski, Wes. *Great Lakes Lighthouses, American and Canadian: A Comprehensive Directory/Guide.* Gwinn, Mich.: Avery Color Studios, 1998.

Ontario Bluewater Visitor Guide. Wellesley, Ontario: Peter Hafemann, 1998.

Ontario Travel Planner. Toronto: Ministry of Economic Development, Trade and Tourism, 1998.

Orr, Blair. "Land Use Change on Michigan's Lake Superior Shoreline: Integrating Land Tenure and Land Cover Type Data." *Journal of Great Lakes Research* 23, no. 3 (September 1997): 328–38.

Ottawa National Wildlife Refuge, Ohio (brochure). Oak Harbor, Ohio: Ottawa National Wildlife Refuge, U.S. Fish and Wildlife Service, n.d.

"The Ottawa River: The Watershed, the Problems, and the Cleanup." *Maumee RAP Newsletter,* June 1998, pp. 1–2.

Our Lakes, Our Health, Our Future: Environmental Presentation Summaries, International Joint Commission Public Forum, Niagara Falls, Ontario, November 1, 1997 (brochure). Buffalo: Great Lakes United, 1997.

Peterson, Rolf O. *Ecological Studies of Wolves on Isle Royale.* Annual Report, 1997–98. Houghton, Mich.: Michigan Technological University, 1998.

———. *The Wolves of Isle Royale: A Broken Balance.* Foreword by L. David Mech. Minocqua, Wis.: Willow Creek Press, 1995.

Petrie, Shelly. "Come to the 1998 Citizen Hearings on Great Lakes—St. Lawrence River Water Pollution." *Great Lakes United* 12, no. 2 (Spring 1998): 32, 23.

Pettibone, Gary W., and Kim N. Irvine. "Levels and Sources of Indicator Bacteria Associated with the Buffalo River 'Area of Concern,' Buffalo, New York." *Journal of Great Lakes Research* 22, no. 4 (December 1996): 896–905.

Pictured Rocks Official Map and Guide (brochure). Munising, Mich.: Pictured Rocks National Lakeshore, National Park Service, n.d.

The Plight of Migratory Birds in Ohio (brochure). Oak Harbor, Ohio: Partners in Flight—Ohio Working Group, n.d.

Point Iroquois Light Station (brochure). Sault Ste. Marie, Mich.: Hiawatha National Forest, U.S.D.A. Forest Service, n.d.

Point Pelee National Park 1998 Visitor Guide. Leamington, Ontario: Point Pelee National Park, Canadian Heritage Parks Canada, 1998.

Porcupine Mountains Wilderness State Park, Ontanagon, Michigan (brochure). Lansing: Michigan Department of Natural Resources, Parks and Recreation Division, 1997.

Potter, Heather. "Our Basin Heritage." *Great Lakes United* 11, no. 2 (Spring 1997): eight-page insert (n.p.).

Program Strategies 1999–2003 for the Cuyahoga River Remedial Action Plan (RAP) Coordinating Committee. Cleveland: Strategic Planning Committee, Cuyahoga River Remedial Action Plan Coordinating Committee, March 25, 1999.

Protecting and Restoring the Great Lakes Ecosystem: Today's Issues and Challenges. Chicago: Great Lakes National Program Office, U.S. Environmental Protection Agency, 1996.

Quinn, Frank H., et al. "Laurentian Great Lakes Hydrology and Lake Levels under the Transposed 1993 Mississippi River Flood Climate." *Journal of Great Lakes Research* 23, no. 3 (September 1997): 317–27.

"Quinte, Bay of." *Encyclopedia Britannica (Micropedia)*, 9:863.

Rafferty, Claudine Jones, and Joel Kaplan. *Public Health Assessment, Pollution Abatement Services (Pas), City of Oswego, Oswego County, New York: Cerclis No. Nyd000511659, May 30, 1997. Prepared by New York State Department of Health under Cooperative Agreement with the Agency for Toxic Substances and Disease Registry.* Agencies for Toxic Substances and Disease Registry Web site (http://atsdr1.atsdr.cdc.gov/HAC/PHA/PAS/pas_toc.html), accessed March 19, 1999.

Rafferty, Michael, and Robert Sprague. *Porcupine Mountains Companion: Inside Michigan's Largest State Park.* 3rd ed. White Pine, Mich.: Nequaket Natural History Associates, 1996.

Rainbow Country Discovery Guide, 1998. Sudbury, Ontario: Rainbow Country Travel Association, 1998.

Rankin, David, and Susan Crispin. *The Conservation of Biological Diversity in the Great Lakes Ecosystem: Issues and Opportunities.* Chicago: The Nature Conservancy Great Lakes Program, 1994.

"Rehabilitating and Conserving Detroit River Habitats: Success Stories." Citizens Environment Alliance of Southwestern Ontario and Southeastern Michigan Web site (http://www.mnsi.net/cea/drhc/success.html), accessed February 24, 1999.

Reid, Ron, and Karen Holland. *The Land by the Lakes: Nearshore Terrestrial Ecosystems.* State of the Lakes Ecosystem Conference 1996 Background Paper. Chicago: U.S. Environmental Protection Agency Great Lakes National Program Office, 1997.

Rennicke, Jeff. *Isle Royale: Moods, Magic, and Mystique.* Houghton, Mich.: Isle Royale Natural History Association, 1989.

Review of Government Resources and Changing Program Thrusts as They Relate to Delivery of Programs under the Great Lakes Water Quality Agreement. Windsor, Ontario: Great Lakes Water Quality Board, International Joint Commission, 1998.

Rives, Lisa. "Cuyahoga River Area of Concern." U.S. EPA Great Lakes Program Office Web site (http://www.epa.gov/glnpo/aoc/cuyahoga.html), accessed March 9, 1999.

Rose, Robert. *Pictured Rocks Resource Report: Overview of Cambrian Sandstone Environments of Deposition* (brochure). Munising, Mich.: Pictured Rocks National Lakeshore, National Park Service, 1998.

The Sault Ste. Marie Canal: A National Historic Site. Sault Ste. Marie, Ontario: Saulte Ste. Marie Canal National Historic Site, Canadian Heritage Parks Canada in conjunction with the *Sault Star,* 1997.

Sault Ste. Marie Canal Attikamek Trail (brochure). Sault Ste. Marie, Ontario: Environment Canada Parks Service, 1991.

Schmidt, Wayne A. "U.S. Feds Rein in Rogue States." *Great Lakes United* 11, no. 2 (Spring 1997): 1–2.

Schoolcraft, Henry Rowe. *Travels through the Northwestern Regions of the United States.* Albany, N.Y.: E & E Hosford, 1821 (facsimile by Readex Microprint, 1966).

Scott, Robert W., and Floyd A. Huff. "Impacts of the Great Lakes on Regional Climate Conditions." *Journal of Great Lakes Research* 22, no. 4 (December 1996): 845–63.

Self-Guided Hike/Bike Tour—South Bass Island. Columbus: Ohio Sea Grant Program, Ohio State University, 1985.

Selkirk Shores State Park (brochure). Pulaski, N.Y.: Selkirk Shores State Park, State of New York Office of Parks, Recreation, and Historic Preservation, n.d.

Singing Sands Almanac (brochure). Porter, Ind.: Indiana Dunes National Lakeshore, National Park Service, 1998.

Sleeping Bear Dunes Official Map and Guide (brochure). Empire, Mich.: Sleeping Bear Dunes National Lakeshore, National Park Service, n.d.

Snyder, Fred. "Is the Ruffe Less of a Threat?" *Littoral Drift* (University of Wisconsin Sea Grant Institute), May/June 1998, p. 2.

"Something for Everyone!" *Tobermory Welcomes You.* Tobermory Chamber of Commerce Web site (http://www.tobermory.org/welcome.htm), accessed March 31, 1999.

Souder, William. "30,000 Logs Under the Sea: Lake Superior Begins Yielding a Historic Trove of Timber." *Washington Post,* August 14, 1996, p. 1.

"Special Section on the ARCS Program," *Journal of Great Lakes Research* 22, no. 3 (September 1996): 493–682.

State of Ohio 1998 State of the Lake Report: Lake Erie Quality Index. Toledo: Ohio Lake Erie Commission, 1998.

State of the Great Lakes 1997: The Year of the Nearshore. Burlington, Ontario, and Chicago: Environment Canada Office of the Regional Science Advisor and U.S. Environmental Protection Agency Great Lakes National Program Office, 1997.

Strutin, Michele. *The Smithsonian Guides to Natural America: The Great Lakes, Ohio–Indiana–Michigan–Wisconsin.* Washington, D.C.: Smithsonian Books, 1996.

Sullivan, Jerry. *Chicago Wilderness: An Atlas of Biodiversity.* Chicago: Chicago Region Biodiversity Council, n.d.

Superior Water-Logged Lumber Company, Inc., Manufacturers of Timeless Timber (brochure). Ashland, Wis.: Superior Water- Logged Lumber Company, Inc., n.d.

"Surface Water Assessment—Lake Michigan." *Illinois Water Quality Report 1992–1993.* Consortium for International Earth Science Information Network Web site (http://epawww.ciesin.org/glreis/nonpo/ndata/IEPA/IWQReport.html), accessed February 1, 1999.

Tettegouche State Park (brochure). St. Paul: State of Minnesota Department of Natural Resources, 1998.

This Is Manitoulin 1998. Little Current, Ontario: The Manitoulin Expositor, 1998.

Thompson, Jon, and Nancy Lemmen. "Experience the Tradition." Weyerhaeuser Au Sable River Canoe Marathon Web site (http://www.ausablecanoemarathon.org/exptra98.html), accessed February 17, 1999.

Thorp, Steve, Ray Rivers, and Victoria Pebbles. *Impacts of Changing Land Use: State of the Lakes Ecosystem Conference (SOLEC) 1996 Working Paper.* Ann Arbor, Mich., and Burlington, Ontario: Great Lakes Commission and Environment Canada, 1996.

Tifft Nature Preserve: Buffalo's Unique Urban Nature Sanctuary (brochure). Buffalo, N.Y.: Buffalo Museum of Science, n.d.

"Tifft Nature Preserve: Our Unique Urban Nature Preserve." Buffalo Museum of Science Web site (http://www.sciencebuff.org/Tifft_Nature_Preserve/), accessed March 16, 1999.

"Tobermory Lighthouses." *Tobermory Welcomes You.* Tobermory Chamber of Commerce Web site (http://www.tobermory.org/lighthouses.htm), accessed March 31, 1999.

"Tonnage for Selected U.S. Ports in 1997 Ranked by Total Tons." *Waterborne Commerce Statistics Center.* U.S. Army Corps of Engineers Water Resources Support Center Web site (http://www.wrsc.usace.army.mil/ndc/wcporton.htm), accessed April 20, 1999.

Townsend, Robert B. "The Carrying Place Road." Prince Edward Radio Club [Picton, Ontario] Web site (http://www.stormy.ca/perc/sqfcarry.html), accessed March 23, 1999.

Tritsch, Shane. "Deep Trouble." *Chicago Magazine,* September 1998, pp. 63–82.

U.S.D.A. Forest Service Eastern Region: Providing Conservation Leadership." U.S.D.A. Forest Service Eastern Region Web site (http://www.fs.fed.us/outernet/r9/welcome.htm), accessed December 18, 1998.

View from the Dunes (brochure). Empire, Mich.: Sleeping Bear Dunes National Lakeshore, National Park Service, 1998.

Visitor Guide, Stone Laboratory: Ohio's Lake Erie Laboratory since 1895. Put-in Bay, Ohio: Franz Theodore Stone Laboratory, Ohio State University, 1998.

Welcome to Door County. Sturgeon Bay, Wis.: Door County Chamber of Commerce, 1998.

White, Erin. *Realizing Remediation: A Summary of Contaminated Sediment Remediation Activities in the Great Lakes Basin.* Chicago: U.S. Environmental Protection Agency Great Lakes National Program Office, 1998.

"Whitefish Dunes State Park Offers Namesake Dunes, Programs, and Trails." Door County Chamber of Commerce Web site (http://doorcountyvacations.com/Parks/White.html), accessed December 17, 1998.

Whitefish Dunes State Park Trail Map (brochure). Sturgeon Bay, Wis.: Friends of Whitefish Dunes, 1996.

Wisconsin's Biodiversity as a Management Issue: A Report to Department of Natural Resources Managers, May 1995. Madison: Wisconsin Department of Natural resources, 1995.

Wooster, Margaret. "Great Lakes United Position Statement on RAPs." *Journal of Great Lakes Research* 23, no. 2 (June 1997): 224.

———. "Why Are the Governments Abandoning RAPs?" *Great Lakes United* 11, no. 2 (Spring 1997): 4.

INDEX

Castle Park, Michigan (*cont.*)
339
Catawba Island (Lake Erie), 170, 172, 185
Cavanaugh's (Chicago restaurant), 97
Cave Point (Door Peninsula), 35–37, 43
Cedar Point, Ohio, 184
Center for Lake Erie Area Research, 159
Central Basin (Lake Erie), 168, 194
Cervidae (deer, elk, and moose), and
forests, 325–27
Chagrin River (Ohio), 207
Champlain, Samuel de (French
explorer), 20, 242
change: acceptance of, 249–50; in
author's biases, 23–24; difficulty in
noticing, 213–14; natural, on Isle
Royale, 327. *See also* camping,
changes in
chaos theory, and wildlife management,
73–74, 75, 193
Charlevoix, Lake (Michigan), 128–29, 130
Charlevoix, Michigan, 128
Chatham, Ontario, 151
Chaumont, New York, 242
Chaumont Barrens (New York alvar),
242–44; *photo, 243*
"Chemical Valley" (Ontario), 150
Chequamegon Bay (Lake Superior), 311,
312; *photo, 313*
Chequamegon National Forest
(Wisconsin), 56–57
Chicago, Illinois, 20, 21, 54–55, 75, 76, 77,
78–79, 86–87, 104; Great Fire, 98,
101; growth of, 20, 81–82; yacht
basin, 98; *photo, 79*
Chicago, Lake (geology), 150
Chicago Region Biodiversity Council,
88–89
Chicago River (Illinois), 78, 84, 87, 102;
cleanup of, 78–79; diversion of, 21,
77–78, 337; geology of, 77; reversal
of (*see* diversion of); *photo, 79*
Chicago Sanitary District, 78
"Chicago Wilderness" (biodiversity
initiative), 88–89, 338
Chi-Cheemaun (Lake Huron ferry),
258–59, 261–63
chinook salmon, in Lake Ontario, 229

Chippewa Nation (Native American
tribe), as casino operators, 281
Chippewa National Forest (Minnesota),
57
Chippewa-Stanley Low-Water Stage
(geology), 20, 42
Chitwood, Larry, 23, 32, 35, 39, 78, 126,
172, 204, 234, 247, 269, 271
chlorine, and paper making, 81, 270
Christmas tree reef, at Collingwood,
Ontario, 254
Chutes, The (Ontario waterfall), 271
clams. *See* unionid mussels
Clark, Milton, 78, 91–95
Clarke, Andrew, 149, 150
Clayton, New York, 244
Clean Air Act, 156
Clean Water Act, 156
CLEAR. *See* Center for Lake Erie Area
Research
Clear Lake (Espanola, Ontario), 267
clearcuts, as wildlife management tool,
137–38
Cleveland, Ohio, 17, 18, 21, 180, 201, 207
Cleveland Heights, Ohio, 201
climate change, and Great Lakes levels,
87–88, 188–89
Clinton River (Michigan), 164
CN Tower (Toronto, Ontario), 234
coal, popular unfamiliarity with, 302
Coast Range, Oregon, compared to the
Minnesota coast range, 317
Cobourg, Ontario, 250
"coffee trick" (boating), 329
coho salmon, in Lake Ontario, 229
Collingwood, Ontario, 65, 161, 251,
252–55, 257; *photo, 253*
Collingwood Stakeholder Group, 254,
255
command-and-control legislation,
problems with, 339
Community-Based Environmental
Planning (Environmental
Protection Agency program), 84
cones of depression (hydrology), and
diversions of Great Lakes water, 67,
337
confined disposal facilities, 55, 180–81

Titles in the Great Lakes Books Series

Freshwater Fury: Yarns and Reminiscences of the Greatest Storm in Inland Navigation, by Frank Barcus, 1986 (reprint)

Call It North Country: The Story of Upper Michigan, by John Bartlow Martin, 1986 (reprint)

The Land of the Crooked Tree, by U. P. Hedrick, 1986 (reprint)

Michigan Place Names, by Walter Romig, 1986 (reprint)

Luke Karamazov, by Conrad Hilberry, 1987

The Late, Great Lakes: An Environmental History, by William Ashworth, 1987 (reprint)

Great Pages of Michigan History from the Detroit Free Press, 1987

Waiting for the Morning Train: An American Boyhood, by Bruce Catton, 1987 (reprint)

Michigan Voices: Our State's History in the Words of the People Who Lived It, compiled and edited by Joe Grimm, 1987

Danny and the Boys, Being Some Legends of Hungry Hollow, by Robert Traver, 1987 (reprint)

Hanging On, or How to Get through a Depression and Enjoy Life, by Edmund G. Love, 1987 (reprint)

The Situation in Flushing, by Edmund G. Love, 1987 (reprint)

A Small Bequest, by Edmund G. Love, 1987 (reprint)

The Saginaw Paul Bunyan, by James Stevens, 1987 (reprint)

The Ambassador Bridge: A Monument to Progress, by Philip P. Mason, 1988

Let the Drum Beat: A History of the Detroit Light Guard, by Stanley D. Solvick, 1988

An Afternoon in Waterloo Park, by Gerald Dumas, 1988 (reprint)

Contemporary Michigan Poetry: Poems from the Third Coast, edited by Michael Delp, Conrad Hilberry and Herbert Scott, 1988

Over the Graves of Horses, by Michael Delp, 1988

Wolf in Sheep's Clothing: The Search for a Child Killer, by Tommy McIntyre, 1988

Copper-Toed Boots, by Marguerite de Angeli, 1989 (reprint)

Detroit Images: Photographs of the Renaissance City, edited by John J. Bukowczyk and Douglas Aikenhead, with Peter Slavcheff, 1989

Hangdog Reef: Poems Sailing the Great Lakes, by Stephen Tudor, 1989

Detroit: City of Race and Class Violence, revised edition, by B. J. Widick, 1989

Deep Woods Frontier: A History of Logging in Northern Michigan, by Theodore J. Karamanski, 1989

Orvie, The Dictator of Dearborn, by David L. Good, 1989

Seasons of Grace: A History of the Catholic Archdiocese of Detroit, by Leslie Woodcock Tentler, 1990

The Pottery of John Foster: Form and Meaning, by Gordon and Elizabeth Orear, 1990

The Diary of Bishop Frederic Baraga: First Bishop of Marquette, Michigan, edited by Regis M. Walling and Rev. N. Daniel Rupp, 1990

Walnut Pickles and Watermelon Cake: A Century of Michigan Cooking, by Larry B. Massie and Priscilla Massie, 1990

The Making of Michigan, 1820–1860: A Pioneer Anthology, edited by Justin L. Kestenbaum, 1990

America's Favorite Homes: A Guide to Popular Early Twentieth-Century Homes, by Robert Schweitzer and Michael W. R. Davis, 1990

Beyond the Model T: The Other Ventures of Henry Ford, by Ford R. Bryan, 1990

Life after the Line, by Josie Kearns, 1990

Michigan Lumbertowns: Lumbermen and Laborers in Saginaw, Bay City, and Muskegon, 1870–1905, by Jeremy W. Kilar, 1990

Detroit Kids Catalog: The Hometown Tourist, by Ellyce Field, 1990

Waiting for the News, by Leo Litwak, 1990 (reprint)

Detroit Perspectives, edited by Wilma Wood Henrickson, 1991

Life on the Great Lakes: A Wheelsman's Story, by Fred W. Dutton, edited by William Donohue Ellis, 1991

Copper Country Journal: The Diary of Schoolmaster Henry Hobart, 1863–1864, by Henry Hobart, edited by Philip P. Mason, 1991

John Jacob Astor: Business and Finance in the Early Republic, by John Denis Haeger, 1991

Survival and Regeneration: Detroit's American Indian Community, by Edmund J. Danziger, Jr., 1991

Steamboats and Sailors of the Great Lakes, by Mark L. Thompson, 1991

Cobb Would Have Caught It: The Golden Age of Baseball in Detroit, by Richard Bak, 1991

Michigan in Literature, by Clarence Andrews, 1992

Under the Influence of Water: Poems, Essays, and Stories, by Michael Delp, 1992

The Country Kitchen, by Della T. Lutes, 1992 (reprint)

The Making of a Mining District: Keweenaw Native Copper 1500–1870, by David J. Krause, 1992

Kids Catalog of Michigan Adventures, by Ellyce Field, 1993

Henry's Lieutenants, by Ford R. Bryan, 1993

Historic Highway Bridges of Michigan, by Charles K. Hyde, 1993

Lake Erie and Lake St. Clair Handbook, by Stanley J. Bolsenga and Charles E. Herndendorf, 1993

Queen of the Lakes, by Mark Thompson, 1994

Iron Fleet: The Great Lakes in World War II, by George J. Joachim, 1994

Turkey Stearnes and the Detroit Stars: The Negro Leagues in Detroit, 1919–1933, by Richard Bak, 1994

Pontiac and the Indian Uprising, by Howard H. Peckham, 1994 (reprint)

Charting the Inland Seas: A History of the U.S. Lake Survey, by Arthur M. Woodford, 1994 (reprint)

Ojibwa Narratives of Charles and Charlotte Kawbawgam and Jacques LePique, 1893–1895. Recorded with Notes by Homer H. Kidder, edited by Arthur P. Bourgeois, 1994, co-published with the Marquette County Historical Society

Strangers and Sojourners: A History of Michigan's Keweenaw Peninsula, by Arthur W. Thurner, 1994

Win Some, Lose Some: G. Mennen Williams and the New Democrats, by Helen Washburn Berthelot, 1995

Sarkis, by Gordon and Elizabeth Orear, 1995

The Northern Lights: Lighthouses of the Upper Great Lakes, by Charles K. Hyde, 1995 (reprint)

Kids Catalog of Michigan Adventures, second edition, by Ellyce Field, 1995

Rumrunning and the Roaring Twenties: Prohibition on the Michigan- Ontario Waterway, by Philip P. Mason, 1995

In the Wilderness with the Red Indians, by E. R. Baierlein, translated by Anita Z. Boldt, edited by Harold W. Moll, 1996

Elmwood Endures: History of a Detroit Cemetery, by Michael Franck, 1996

Master of Precision: Henry M. Leland, by Mrs. Wilfred C. Leland with Minnie Dubbs Millbrook, 1996 (reprint)

Haul-Out: New and Selected Poems, by Stephen Tudor, 1996

Kids Catalog of Michigan Adventures, third edition, by Ellyce Field, 1997

Beyond the Model T: The Other Ventures of Henry Ford, revised edition, by Ford R. Bryan, 1997

Young Henry Ford: A Picture History of the First Forty Years, by Sidney Olson, 1997 (reprint)

The Coast of Nowhere: Meditations on Rivers, Lakes and Streams, by Michael Delp, 1997

From Saginaw Valley to Tin Pan Alley: Saginaw's Contribution to American Popular Music, 1890–1955, by R. Grant Smith, 1998

The Long Winter Ends, by Newton G. Thomas, 1998 (reprint)

Bridging the River of Hatred: The Pioneering Efforts of Detroit Police Commissioner George Edwards, by Mary M. Stolberg, 1998

Toast of the Town: The Life and Times of Sunnie Wilson, by Sunnie Wilson with John Cohassey, 1998

These Men Have Seen Hard Service: The First Michigan Sharpshooters in the Civil War, by Raymond J. Herek, 1998

A Place for Summer: One Hundred Years at Michigan and Trumbull, by Richard Bak, 1998

Early Midwestern Travel Narratives: An Annotated Bibliography, 1634–1850, by Robert R. Hubach, 1998 (reprint)

All-American Anarchist: Joseph A. Labadie and the Labor Movement, by Carlotta R. Anderson, 1998

Michigan in the Novel, 1816–1996: An Annotated Bibliography, by Robert Beasecker, 1998

"Time by Moments Steals Away": The 1848 Journal of Ruth Douglass, by Robert L. Root, Jr., 1998

The Detroit Tigers: A Pictorial Celebration of the Greatest Players and Moments in Tigers' History, updated edition, by William M. Anderson, 1999

Father Abraham's Children: Michigan Episodes in the Civil War, by Frank B. Woodford, 1999 (reprint)

Letter from Washington, 1863–1865, by Lois Bryan Adams, edited and with an introduction by Evelyn Leasher, 1999

Wonderful Power: The Story of Ancient Copper Working in the Lake Superior Basin, by Susan R. Martin, 1999

A Sailor's Logbook: A Season aboard Great Lakes Freighters, by Mark L. Thompson, 1999

Huron: The Seasons of a Great Lake, by Napier Shelton, 1999

Tin Stackers: The History of the Pittsburgh Steamship Company, by Al Miller, 1999

Art in Detroit Public Places, revised edition, text by Dennis Nawrocki, photographs by David Clements, 1999